Critical Realism and the Objective Value of Sustainability

Critical Realism and the Objective Value of Sustainability contributes to the growing discussion surrounding the concept of sustainability, using a critical realist approach within a transdisciplinary theoretical framework to examine how sustainability objectively occurs in the natural world and in society.

The book develops an ethical theory of sustainability as an objective value, rooted not in humans' subjective preferences but in the holistic web of relationships, interdependencies, and obligations existing among living things on Earth, a web believed to have maintained life on Earth over the last 3.7 billion years. It proposes three pillars of sustainability ethics: contentment for the human existence given to us; justice (beyond distributive justice); and meaningful freedom (within ecological and moral limits). Using abductive reasoning, the book infers that there is an out-of-this-world Sustainer behind the Earth's sustainability acting as a metaphysical source of all being and value. It argues that sustainability value, accepted as a shared understanding of the common good, must guide individual decisions and socio-economic development efforts as a matter of deliberate choice, as well as be built on the awareness that there are non-negotiable, pre-established conditions for our planet's sustainability.

This book will be of interest to students and scholars across fields of inquiry, including sustainability, sustainable development, environmental philosophy and ethics, philosophy of science, and ecological economics, and to whoever may wonder why seasons exists and why humans have creative minds.

Gabriela-Lucia Sabau is Honorary Research Professor of economics and environmental studies at Memorial University of Newfoundland (MUN), Canada. Her research interests are sustainability and sustainable development fuelled by scientific knowledge and value judgements; unjust un-economic growth; demand-side management of water resources; sustainable management of fisheries with a focus on small-scale fisheries; and assessing the potential for agro-ecological transition in Newfoundland and Labrador's agriculture. Gabriela co-founded MUN Grenfell Campus's first PhD program in Transdisciplinary Sustainability (2019) and continues to derive inspiration from the students she supervises in the program.

Routledge Environmental Ethics

Series Editor: Benjamin Hale, University of Colorado, Boulder

The Routledge Environmental Ethics series aims to gather novel work on questions that fall at the intersection of the normative and the practical, with an eye toward conceptual issues that bear on environmental policy and environmental science. Recognizing the growing need for input from academic philosophers and political theorists in the broader environmental discourse, but also acknowledging that moral responsibilities for environmental alteration cannot be understood without rooting themselves in the practical and descriptive details, this series aims to unify contributions from within the environmental literature.

Books in this series can cover topics in a range of environmental contexts, including individual responsibility for climate change, conceptual matters affecting climate policy, the moral underpinnings of endangered species protection, complications facing wildlife management, the nature of extinction, the ethics of reintroduction and assisted migration, reparative responsibilities to restore, among many others.

Philosophy and the Climate Crisis
How the Past can Save the Present
Byron Williston

Climate Justice Beyond the State
Lachlan Umbers and Jeremy Moss

The Concept of Milieu in Environmental Ethics
Individual Responsibility within an Interconnected World
Laÿna Droz

Critical Realism and the Objective Value of Sustainability
Philosophical and Ethical Approaches
Gabriela-Lucia Sabau

For more information on the series, please visit https://www.routledge.com/Routledge-Environmental-Ethics/book-series/ENVE

Critical Realism and the Objective Value of Sustainability
Philosophical and Ethical Approaches

Gabriela-Lucia Sabau

Routledge
Taylor & Francis Group
LONDON AND NEW YORK

earthscan
from Routledge

First published 2024
by Routledge
4 Park Square, Milton Park, Abingdon, Oxon OX14 4RN

and by Routledge
605 Third Avenue, New York, NY 10158

Routledge is an imprint of the Taylor & Francis Group, an informa business

British Library Cataloguing-in-Publication Data
A catalogue record for this book is available from the British Library

Library of Congress Cataloging-in-Publication Data
Names: Sabau, Gabriela-Lucia, author.
Title: Critical realism and the objective value of sustainability :
philosophical and ethical approaches / Gabriela-Lucia Sabau.
Description: Abingdon, Oxon ; New York, NY : Routledge, 2024. |
Series: Routledge environmental ethics | Includes bibliographical references
and index.
Identifiers: LCCN 2023044828 (print) | LCCN 2023044829 (ebook) |
ISBN 9781032310015 (hardback) | ISBN 9781032310022 (paperback) |
ISBN 9781003307587 (ebook)
Subjects: LCSH: Sustainability--Philosophy. | Sustainability--Moral and
ethical aspects. | Environmental ethics.
Classification: LCC GE195 .S225 2024 (print) | LCC GE195 (ebook) |
DDC 179/.1--dc23/eng/20231025
LC record available at https://lccn.loc.gov/2023044828
LC ebook record available at https://lccn.loc.gov/2023044829

ISBN: 978-1-032-31001-5 (hbk)
ISBN: 978-1-032-31002-2 (pbk)
ISBN: 978-1-003-30758-7 (ebk)

DOI: 10.4324/9781003307587

Typeset in Times New Roman
by Taylor & Francis Books

This book is lovingly dedicated to my four wonderful grandchildren: Timothy, Michael, Teofil, and Cassian. Thank you for showing me that sustainability is real and worth accepting as a gift.

Contents

Figures

Acknowledgements

This book was years in the making and would not have been possible without the constant support and encouragement I have generously received from many people. I first became interested in sustainability when working on my PhD degree in the late 1990s. I then discovered with delight the writings of 'giant' thinkers such as Karl Popper, Nicholas Georgescu-Roegen, C.S. Lewis, G.K. Chesterton, and Michael and Karl Polanyi on whose shoulders I am trying to stand. I was also very grateful to work in a team of enthusiastic colleagues led by the late Professor Beniamin Cotigaru on a project for the sustainable reconstruction of Romania. Among the colleagues I remember with thankfulness from that period are Theodor Purcarea, Emilian Dobrescu, Paul Ghita Tanase, Coralia Angelescu, and Christina Marta Suciu.

When I moved to Canada, I introduced ecological economics in my teaching and benefitted from the wisdom of my ecological economist colleagues such as the late Professor Herman Daly, Robert Costanza, Joshua Farley, Jon Erickson, and Brian Czech, from whom I learned that 'crude' GDP growth does not measure sustainability when the environment is being destroyed and social inequality rises. Consciously decided limits to economic activity are needed based on the laws of thermodynamics and on valuation of what is meaningful, namely life and unity of life in biological and social communities, plus human wisdom to achieve it all. I was able to verify the truth of these findings in a beautiful coastal and marine environment during 2008 to 2018 when I was invited to teach a Master's course at the University Centre of the Westfjords in North-Western Iceland. Thank you, Peter Weiss, Sigridur Olafsdottir, Dagny Arnarsdottir, Gunna Sigga, Pernilla Rein, Catherine Chambers, and all the cohorts of students along the way for the opportunity to teach and learn about how achieving fisheries sustainability in an island context requires continuously balancing policies which start from a deep knowledge of the dynamic socio-ecological system. A big thank you to my friend Ratana Chuenpagdee who initiated the Too Big to Ignore global partnership for small-scale fisheries research and invited me to focus my research between 2012 and 2018 on the sustainability of small-scale fisheries worldwide.

Since 2017, I co-developed Memorial University of Newfoundland, Grenfell Campus's PhD in Transdisciplinary Sustainability, which started accepting students in April 2019, and what a ride it has been since! Thank you, Wade Bowers, Kelly Vodden, Lakshman Galagedara, Mano Krishna-pillai, Mumtaz Cheema, Olga Vasilyeva, Garrett Richards, and Michele Piercey-Normore for your hard work and valuable insights. Since the start of the program, I have been inspired by my graduate students to think about the deep questions of life on earth, questions which belong to sustainability even if science alone cannot answer them. Thank you, Temi, Nathaniel, Selim, Naznin, Tithy, Kamal, Natasha, and Sara. Thanks are also due to colleagues and friends who have read parts of the manuscript and provided valuable comments, such as Dumitru Deac, Sean McGrath, John Dagevos, and Rose-lyne Okech. The manuscript has also been carefully read by loving family members. Thank you, Dan and Megan, for giving me the time to write, and for taking time to provide insights and questions which kept improving my writing. Thank you, Dan, also for your expert technical support.

At Routledge, I have been kindly and expertly guided by my editors, Grace Harrison and Matt Shobbrook. Thank you both for believing in this project and for making it happen. Also, thank you, Katherine Laidler, for your hard and excellent work of copyediting the manuscript.

Finally, I am grateful to the One who made the universe and me within it, and who keeps it functioning in a sustainable manner. If this work is of any value, it is not my merit, as I have never accomplished anything meaningful apart from Him who lights my path. I, of course, do accept responsibility for any errors of understanding or interpretation.

Introduction

Sustainability, planet Earth's ability to persist and provide a lasting, intricate life-support system for all living things, is wired in the natural world as a matter of fact. That is made evident in the order and interconnectedness we see in nature, where "all animals and plants are held in a delicate balance, and every entity has its purpose and role in its ecosystem" (Wohlleben, 2017). Sustainability transpires in the yearly succession of seasons and the action of the laws of thermodynamics:

> History discloses two main tendencies in the course of events. One tendency is exemplified in the slow decay of physical nature. [...] The other tendency is exemplified by the yearly renewal of nature in the spring, and by the upward course of biological evolution.
>
> (Whitehead, 1929)

It is obvious in the perpetuation of life over millennia, or in the continuous cycle of life followed by death, with "life persisting in the midst of its perpetual perishing" (Rolston, 2007). Nature exists, majestic and mysterious, as a "great interlocking event in space and time" (Lewis, 1947/2001). It is, yet, open to be discovered, displaying beauty in sights, sounds, smells, shapes, and colours, and offering bountiful wealth. Natural phenomena, which humans call natural disasters, such as landslides, earthquakes, tornadoes, forest fires, and storms, are also part of a dynamic, functioning planet. The nature-aggregate also displays resiliency, or an ability to bounce back after disturbances, and potential for growth, as well as invisible processes, such as carbon cycling, photosynthesis, or the immunity of multicellular organisms. While nature functions on its own apparently, it is obvious that it obeys certain objective rules and limitations which have been pre-established by "forces we do not create and cannot fully control, forces that bring us into being and sustain us and life around us, but forces that also limit and destroy us and determine the destiny of the cosmos" (Gustafson, 1994). Science keeps discovering such pre-established rules (Georgescu-Roegen, 1971), as well as boundaries and ecological limits which have now been exceeded by humankind (Meadows et al., 1972; Rockström et al., 2009; Raworth, 2017).

DOI: 10.4324/9781003307587-1

However, science cannot provide the full picture of nature's sustainability. It can give good descriptions of the parts involved, but not of the synergistic whole which is greater than the sum of its parts. Things become even more complicated when we try to understand how nature's sustainability is impacted by human interventions aiming to either extract things from nature or dispose of waste in the terrestrial or the ocean environment.

There have been successful attempts to model sustainability in complex, "linked social-ecological systems" (Ostrom, 2007), such as forests and fisheries, starting from the idea that humans are a part of nature and they are able to know as well as comply with the objective rules of nature's sustainability. Assuming that the sustainability we see in nature is a good thing, both for nature and for humans, reflective individuals have worked together to identify rules and variables able to lead to the sustainability of these coupled systems by balancing existing resources with identified needs (Ostrom, 2009). Yet, these models do not inform the way most of the modern socio-economic systems are organized and function. Since the age of Enlightenment, from the late seventeenth to the early nineteenth century, and as the Industrial Revolution was unfolding, humans have developed their own definition of sustainability based on a narrative which has become dominant in the Global North. This narrative elevates humans and their needs above life-sustaining nature as if "man has long forgotten that the earth was given to him for usufruct alone, not for consumption, still less for profligate waste" (Marsh, 1965). In this narrative, nature is no longer seen as a complex, dynamic, enveloping, and providing whole, run by forces that humans do not control, but it is reduced to one sub-system among other human-created systems, be they economic, social, or cultural. It was hoped that this sub-system, called ecological or environmental, could somehow be managed to continue providing for the needs of successive generations of humans to continue to exist, if only we could figure out a way to harmoniously integrate it with human-made economic and social systems. This created a purely anthropocentric concept of sustainability.

The term "sustainability" was coined by a German forester, Hans Carl von Carlowitz, in his 1713 book *Sylvicultura Oeconomica* in which he proposed rules for "continuous, permanent and sustainable use" of the forest. These rules recognized humans' dependence on nature and implied deliberate care to conserve nature's ecological integrity as structure and functions in order for the standing forest to continue to serve human communities in perpetuity. The concept implied a balance between new-growth trees and dying ones, as von Carlowitz considered that it is "anything but sensible to cut down more wood in the forest than grows back" (Endres, 2011). The anthropocentric "sustainability" concept was rebranded in 1987 as "sustainable development" by the Brundtland report of the World Commission for Environment and Development. The report defined sustainable development as harmonious global economic, social and ecological development that "meets the needs of the present without compromising the ability of future generations to meet their own needs" (WCED, 1987).

The concept mentions the "essential needs of the world's poor" and some "limits", but defines them in a reductionist manner, as "limitations imposed by the state of technology and social organization on the environment's ability to meet present and future needs", and not as objective limits imposed by nature. In this definition, economic and social development fuelled by "growth and accumulation" (Gowdy, 2014) have been prioritized over the more holistic sustainability of nature, using Robert Solow's theory that nature's capital can be easily substituted by human-made capital (Solow, 1974). The assumption was that continued economic growth, measured as the rate of increase in real gross domestic product (GDP), backed by science and technological progress, will enable humans to solve the problems of both ecological and social sustainability. The anthropocentric "sustainable development" concept has thus swallowed whole the ontologically objective "sustainability" concept, rendering the limits and constraints imposed by nature on its users completely invisible, and making "the material demands of the human species [as] the primary test of what should be done with the Earth" (Du Pisani, 2006). As a consequence, economic growth has become the main ecologically expansive tool of sustainable development as "our civilization has been built on a continuously expanding human environmental footprint [...] a mode of economic expansion that has generated increased material affluence at the cost of growing environmental harm" (Meadowcroft, 2017). It was believed that universal prosperity is achievable and desirable based on "the dominant modern belief [is] that the soundest foundation of peace would be universal prosperity" (Schumacher, 1973). The practical application of the "sustainable development" concept in the last almost four decades has led humanity on an unsustainable path, manifested in deep and protracted environmental and social crises, while "only limited and halting steps are being taken to secure Homo Sapiens a liveable future" (Gale, 2018). Indeed, assessments of the global environmental outlook show that:

> Current methods of generating material prosperity have undermined ecosystem health and caused massive environmental damage, crossing several of these planetary boundaries, to the point where the development of human societies and the "safe operating space" for human life on Earth is at risk.
>
> (UN Environment, 2019)

Another assessment identifies humans as irreversibly threatening the "unusual" stability of the planet's environment over the past 10,000 years which has made "human civilizations arise, develop and strive" (Rockström et al., 2009). Some scientists even believe that we have entered a new geological epoch, called the Anthropocene (Crutzen, 2006), an epoch in which humans are the most important geomorphic agent on the planet (Wilkinson and McElroy, 2007). In 2019, the Sustainable Development report noted that "our goal to end extreme poverty by 2030 is being jeopardized as we struggle to respond to entrenched deprivation, violent conflicts and vulnerabilities to natural disasters" (UN, 2019). The COVID-19 pandemic has produced a global recession threatening to reverse into poverty hundreds of millions of people in the developing world (World Bank, 2020). The pandemic has high-lighted a serious acceleration of some radical changes determined by two factors: (i) the Industrial Revolution 4.0 (Schwab, 2016), in particular its core component relating to disruptive technologies; and (ii) climate change and related environmental issues. These radical changes have an all-embracing character ranging from the economy to health, education and, generally, all aspects of society, including what basic values we must adopt. We can see and feel the changes in daily life.

Big-data technology, the internet of things, advanced robotics, and, more recently, the mainstreaming of artificial intelligence (AI) are changing the way humans are to operate. Similarly, climate change is deeply impacting the whole spectrum of human activity, as we begin to already experience a first-wave of environmentally induced relocations, changes in weather patterns, an acceleration of desertification, water shortages, and significant biodiversity loss. The entire configuration of the planet is under pressure, including far-reaching implications in terms of both geopolitics and geoeconomics.

Assuming that this brief assessment is correct, the question arises with respect to the management of these two phenomena. On the one hand, we can note that there is a positive interplay between the two: better and more efficient technologies may assist with a better use of renewable energies. For instance, the internet of things may have positive implications for water management or increased efficiency in the use of energy. Yet, there are serious negative implications, as the ethical underpinnings of these new evolutions are not as readily up for discussion or debate. Scientists have yet to reach a con-sensus on, for example, the environmental impact of electric car batteries, or on humans' ability to control AI's long-term impact on human life. How, then, are we to maximize synergies between these two phenomena and mini-mize negatives? Is it possible to mitigate their adverse effects on society, including its moral values? How can we amend and enrich a concerted action by policy makers, civil society, and the business sector not only in the gov-ernance of each phenomenon but also of the interchange between them? Deeper questions arise: will we be able to control these powerful changes, or will the future witness a runaway project crushing our civilization on the planet which was wired for sustainability?

The purpose of this book is not to inventory the developments in these two areas of change, nor to identify actions taken to date. Instead, I aim to clarify the confusion between the concepts of "sustainability" and "sustainable development", as well as identify a cross-cutting understanding that may allow humans to respond to the current moment; there are challenges and opportunities that arise from the multifaceted dimensions of both disruptive technologies and climate change that impact sustainability. These must be dealt with within the bounds of our well-ordered and miraculously fine-tuned biosphere that we have inherited. The book is therefore trying to testify and exemplify how the concept of sustainability is worth updating and reshaping in order to enable humans to meet both old and new challenges, including of the intertwined phenomena indicated above.

Sustainability is not a new concept. Elements of it can be found many centuries back. In the 1970s, the Club of Rome and then the Brundtland Commission have structured and detailed the concept, which was included in internationally negotiated documents (the United Nations Conference on Sustainable Development) (UN, 1992, 2012, 2019). Unfortunately, since then it has become an empty concept despite the fact that it continues to be referred to. Even more important, its objective ecological dimension has disappeared from centre stage, cementing humans' current stance as powerful rulers of nature rather than humble stewards interested in the long-term survival of the biosphere and its inhabitants.

The book argues for a holistic reconsideration of sustainability which exists not only as a theoretical concept but also as an objective feature of the world, a numinous condition that makes life on planet Earth possible for this and future generations, as the "biospheric membrane that covers the Earth, and you and me [...], is the miracle we have been given" (Wilson, 2002). This view of sustainability implies that the world functions as a web of dynamic inter-relationships in which all living things have specific roles to play, and this is what makes it valuable. However, when humans, as the only creatures endowed with reason and moral impulses, ignore this make-up of the world and their role in it, the life relationship web is broken, and sustainability becomes impossible to achieve. Following Karl Popper (1972/1979, 1982, 1990), this book sees the world as real, objective, comprehensible, and full of possibilities (propensities), and considers that science's goal is to discover truthful and testable causal explanations of how the world functions, aiming to come "nearer to the truth", ideally reaching "absolute truth" (Popper, 1982) or "fundamental truth" (Nabel, 2009). Truth here is understood as correspondence of statements to the facts of the real world, or as a "necessary reality" (Axe, 2016).

> To say that a statement is true is to say that it can correctly be asserted, that it is *such that* it can perform its cognitive function, that it will express facts about the referent, that it will disclose, or give knowledge about, the referent, that it will tell us what is, or was, the case [...] Hence to assert *p* involves the belief that it is true, that is, gives factual knowledge. If *p*, "It is raining," is true, then it *is* raining.
>
> (Sellars, 1946)

A caveat, though, is in order, in the sense that there are truths that science cannot explain. For instance, we "do not know why mathematics is true and whether it is certain" (Berlinski, 2023). That does not mean that mathematicians should stop their efforts to represent the material universe in mathematical formulas. The same is true for sustainability; we might never know why sustainability is true and whether it is certain. However, searching for sustainability with an open mind and multiple tools, and being ready to be amazed in the process, remains a worthwhile endeavour. The book uses a transdisciplinary theoretical framework (Pohl et al., 2017) to develop a critical realist theory of sustainability as it exists both in nature and in the social realm. The transdisciplinary approach uses Ekardt's insight that transdisciplinarity is not "a matter of placing something 'alongside' the classical disciplines and their discourses; rather, it is a matter of their analysis, criticism and continuation" (Ekardt, 2020). This implies the ability to be critical of "scientism" and to accept that there is "deep meaning, truth, relevance, and insight in non-scientific studies pursued with intelligence and rigor" (Hutchinson, 2007) and to accept that intuition, faith, or imagination can be important sources of knowledge. The critical realist perspective is inspired by Roy Bhaskar's critical realism thinking (Bhaskar, 1975, 1979, 2020). For Bhaskar, reality and the human self are complex and dynamic entities, stratified in three fields – the empirical, the actual, and the real – with the real encompassing both the actual and the empirical fields, but also including various powers (natural mechanisms) and their potentialities. Understanding this complex and dynamic reality requires broader methodologies than those offered by scientific materialism, ones able to explain not only the objective facts of the natural world but also to probe into metaphysical hypotheses. The book discusses sustainability both as objective facts of the material world as well as an objective value rooted in the holistic web of relationships existing between living things on planet Earth. Restoring sustainability value, defined and accepted as a shared understanding of the common good of life, both in ecological communities and in social communities, may be a possible solution to the current unsustainable practices based on the subjective values of utilitarianism, pragmatism, and postmodern thinking. Thus understood, sustainability must inform all decisions regarding humans' relations with nature and with each other and their socio-economic development plans. At the individual level, the book develops a sustainability ethics rooted in specific virtues that can change human behaviour towards sustainability. This sustainability ethics has always been embraced by a minority among humans – some Indigenous groups as well as some faith-inspired groups of people living predominantly in premodern cultures. However, recent studies show that transcendental values conducive to sustainable choices and environment conservation are also held by people in Western cultures (Raymond and Kenter, 2016). The book assumes that a primary justice notion is embedded in sustainability, as the life-sustaining system is available without discrimination to all living earthly creatures. A chapter is dedicated to discussing sustainability as justice, beyond the distributive justice system instituted in liberal democracies following the model developed by John Rawls in his *A Theory of Justice* (1971). The book also devotes a chapter to a discussion about sustainability and liberty, understood, *apud* Isaiah Berlin, as both negative and positive liberty, arguing that both types of freedom are important for setting us free from the "evil enchantment of worldliness which has been laid upon us" (Lewis, 1949/2001) for almost 200 years. The book ends with an exploration of the mystery behind the powerful entity that runs the life-support network existing on our life-friendly planet. Here I tread cautiously and with humility, aware that humans might never accurately perceive the whole truth about the Creator and Sustainer of the marvellous life-support system existing on Earth, but I feel it is my duty to investigate the "universal design intuition" (Axe, 2016) of the awesomeness of life with the tools available to me, both scientific and metaphysical.

The book is organized as follows.

The Introduction broaches the topic and explains the rationale of the book, clarifying the confusion around the two concepts "sustainability" and "sustainable development" and identifying a knowledge gap in the sustainability debate, where the ethical aspects are insufficiently discussed. A sustainability ethics framework is proposed, as a possible solution to humanity's current unsustainable way of life. The book argues that sustainability implementation requires a change of mind and a change of heart at the individual self level, implying a new attitude towards nature and towards other living beings, including other humans.

Chapter 1 offers a brief history of sustainability as a concept, as well as its evolution from the notion that humans need to live in harmony with nature as they organize their socio-economic lives, to a notion that they need to also live in harmony with each other. Various definitions of sustainability are discussed, including those embedded in the three-legged concept of "sustainable development". The chapter points to the basic differences between "sustainability" and "sustainable development" and proposes a new definition of sustainability starting from the idea that sustainability is not only a theoretical concept but exists objectively embedded in the make-up of the Universe.

Chapter 2 explains the choice of critical realism and of its richer research methodologies, which include abductive and retroductive inferences, in our search to understand sustainability as an objective feature of the world, which has both visible and invisible aspects. It assumes that the world exists "out there", it is complex, stratified (Bhaskar, 1975), dynamic and open, a "world of propensities" (Popper, 1990). Being comprehensible, it can be studied and understood not only in its visible aspects but also in its invisible characteristics which can reveal effects and causes hard to perceive empirically, but which, however, can be tested by using the "method of hypothesis" (Peirce, 1960).

Chapter 3 discusses sustainability as objective physical reality on our planet, miraculously well-ordered and fine-tuned for supporting life, but at the same time characterized by uncertainty in our continuously "self-arising" biosphere (Bonnett, 2015). The goal of this chapter is to show, based on scientific evidence from physics, astronomy, chemistry, cosmology, and biology, that sustainability exists objectively in nature and is not the result of blind forces, as it is embodied in structures, functions, and processes that seem specifically designed to protect and sustain life. These structures, functions, and processes can be investigated by science through a spiralling process of discovery which can take us closer to understanding the essence of sustainability in nature. On this view, "science is not an epiphenomenon of nature, nor is nature a product of man" (Bhaskar, 1975). While the chapter refers mostly to sustainability in the macrocosm, it also contains some references to the mystery of life in the quantum world which are relevant for sustainability. Understanding quantum fields, waves and particles, and entanglement phenomena is essential for understanding how our sustainable world, made of matter, energy, and "forces", operates, and it can even give us clues about the Sustainer.

Chapter 4 investigates sustainability as it exists objectively in the social world, and asks the question why humans cannot, as the only members of the biotic community endowed with reason and moral impulses, make sustainable choices. The answer proposed considers three causes: humans' limited understanding of reality, science's reductionist view of human nature, and the socio-cultural-political impact of the modernity narrative, which has given rise to a dogmatic universal theory of how the world functions and how the autonomous modern self relates to it. These causes and the impact of postmodern thinking have produced a fragmented sustainability narrative leading to confusion, disunity, and anxiety, ripping up the very fabric of civilization in the Global North. Globalization and the continued technology disruption, idealized and perpetuated by the Silicon Valley "move fast and break things" culture, have contributed to the spreading of this fragmented narrative, deepening social inequalities and potentially irreversibly damaging the environment. The critical realist Transformational Model of Social Activity (Bhaskar, 1975, 1979) is proposed as a potential way to bring change toward sustainability at the individual level. The model implies opening our culture to a process of staged transition, based on a new sustainability narrative in which humans are seen acting as responsible agents of change.

Chapter 5 deals with the ethical dimension of sustainability. It argues for ethical objectivism and for virtue ethics in analyzing sustainability. The intrinsic value of sustainability is discussed under three aspects: as persistence, or temporal and spatial continuity of the life-support system; as goodness of life as a unified whole; and as unity of the biotic community, or the relational character of sustainability. As such, sustainability should be a non-negotiable normative goal overarching all human decisions. However, sustainability is ignored by most individuals in the liberal democracies functioning according to the postmodern narrative built around an artificial, fragmented, individualistic, and self-interested human self, unable to conceive of and live a sustainable life. The chapter develops a virtue-based ethical framework of a sustainable person's identity, and discusses the transformation process through which a modern egotistic individual can become a sustainable person.

Chapter 6 entitled *Sustainability as Justice: Beyond Distributive Justice* discusses the justice aspects of sustainability, starting from the idea that a basic principle of justice is embedded in the sustainability order existing on planet Earth, which protects all life equally. Noting that currently the sustainable development concept only refers to intergenerational justice as a normative requirement, and that intragenerational justice normally administered by liberal democracies using John Rawls' theory of "fair" distributive justice (Rawls, 1971) is inadequate to "right" social inequalities and the overuse of natural resources, a new understanding of sustainability as justice is proposed, as primary and reactive justice rooted in the human right to life.

Chapter 7 discusses the relevance of freedom to understanding and implementing sustainability. Starting from Isaiah Berlin's conception of negative and positive freedom (Berlin, 1969), I argue that while freedom is the objective birthright of any human being, the choice of a sustainable lifestyle depends on understanding and acting upon both aspects of liberty, negative and positive, which sometimes require deliberate sacrifices of freedom. The chapter presents the history of liberty protection under the classical and the modern liberalist doctrines, and the birth of the utilitarian doctrine of moral individualism which instituted freedom as preference satisfaction. Three causes of erosion of individual freedom under modern liberalism are discussed, namely seeing humans as "homo economicus", the creation of a free-market society, and the fast process of producing and spreading the Silicon Valley culture and the tech solutionism approach culminating in the current efforts to achieve AGI (artificial generalized information) systems.

Chapter 8 entitled *Who Is the Sustainer?* argues for the need to reconsider metaphysics as an important source of knowledge about reality. It uses the stratified reality concept of critical realism (Bhaskar, 2020) and the method of inductive reasoning called inference to the best explanation (Lipton, 1991) to search for the most plausible cause of the life-sustaining web of relationships in which humans live. Comparing the worldview of naturalism or scientific materialism with the worldview of theism and the existing evidence about the origin of the Universe, the fine-tuning of the Universe to support life, and the facts provided by the genome analysis in biology, proving the complex design in life-supporting protein molecules, it posits that theism offers a better explanation about the "out-of-this-world" power behind the complex, dynamic, and diverse life-sustaining network than scientific materialism.

Chapter 9 concludes. This book's main message is a message of cautious hope. Sustainability is a magnificent life project, intelligibly conceived to secure existence and continuation of life on our planet. Of all the billions of creatures that live on Earth, only humans have been offered the choice to participate as a force for good in the sustainability project, by acknowledging their creature status and willingness to respect all forms of life and to live within the biosphere's limits. Yet, humans have chosen to behave as autonomous rulers of nature, living according to their own poorly conceived sustainability rules, oblivious of life's complexity, diversity, and awesomeness, and effectively indifferent to the pain, suffering, and destruction this can produce. Awareness is rising that the unsustainability model practiced by the autonomous, individualistic humans over the past 200 years in liberal democracies of the Global North is not working, and that the sustainability narrative and practice need revision. This book can contribute to that radical change in both narrative and practice.

References

Axe, D. (2016) *Undeniable How Biology Confirms Our Intuition That Life Is Designed.* New York: HarperOne.

Berlin, I. (1969) "Two Concepts of Liberty." In *Four Essays on Liberty.* Oxford: Oxford University Press, pp. 118–172.

Berlinski, D. (2023) *Science After Babel.* Seattle, WA: Discovery Institute Press.

Bhaskar, R. (1975) *A Realist Theory of Science.* London: Routledge.

Bhaskar, R. (1979) *The Possibility of Naturalism: A Philosophical Critique of the Contemporary Human Sciences.* Atlantic Highlands, NJ: Humanities Press.

Bhaskar, R. (2020) "Critical realism and the ontology of persons." *Journal of Critical Realism* 19(2): 113–120. https://doi.org/10.1080/14767430.2020.1734736.

Bonnett, M. (2015) "Sustainability, the metaphysics of mastery and transcendent nature." In *Sustainability Key Issues,* edited by H. Kopnina and E. Shoreman-Ouimet. London and New York: Routledge.

Crutzen, P.J. (2006) "The Anthropocene." In *Earth System Science in the Anthropocene,* edited by E. Ehlers and T. Kraft. Berlin: Springer, pp. 13–18.

Du Pisani, J.A. (2006) "Sustainable development – historical roots of the concept." *Environmental Sciences,* 3(2): 83–96.

Ekardt, F. (2020) *Sustainability Transformation, Governance, Ethics, Law.* Cham, Switzerland: Springer.

Endres, A. (2011) *Environmental Economics Theory and Policy.* New York: Cambridge University Press.

Gale, F.P. (2018) *The Political Economy of Sustainability.* Cheltenham, UK and Northampton, MA: Edward Elgar.

Gallie, W.B. (1955) "Essentially contested concepts." *Proceedings of the Aristotelian Society,* 56, 167–198. www.jstor.org/stable/4544562.

Georgescu-Roegen, N. (1971) *The Entropy Law and the Economic Process.* Cambridge, MA: Harvard University Press.

Gowdy, J. (2014) "Governance, sustainability and evolution." In *State of the World 2014: Governing for sustainability,* edited by L. Mastny. Washington, DC: Island Press, pp. 31–40.

Gustafson, J.M. (1994) *A Sense of the Divine: The Natural Environment from a Theocentric Perspective.* Cleveland, OH: Pilgrim Press.

Hutchinson, I.H. (2007) "Warfare and Wedlock—Redeeming the Faith-Science Relationship." *Perspectives on Science and Christian Faith,* 59(2): 91–101.

Lewis, C.S. (2001) *Miracles: A Preliminary Study.* New York: HarperOne. (Original work published 1947)

Lewis, C.S. (2001 [1949]) *The Weight of Glory and Other Addresses.* New York: HarperOne.

Lipton, P. (1991) *Inference to the Best Explanation.* London: Routledge.

Marsh, G.P. (1965) *Man and Nature,* edited by D. Lowenthal. Cambridge, MA: Belknap Press.

Meadowcroft, J. (2017) "Sustainable development, limits and growth: reflections on the conundrum." In *Handbook on Growth and Sustainability,* edited by P.A. Victor and B. Dolter. Cheltenham, UK: Edward Elgar: 38-59.

Meadows, D.H., Meadows, D.L., Randers, J., and Behrens, W.W. III (1972) *Limits to Growth.* Falls Church, VA: Potomac Associates/Universe Books.

Nabel, G.J. (2009) "The coordinates of truth." *Science*, 326(5949): 53–54. https://doi.org/10.1126/science.1177637.

Niebhur, H.R. (1952) "The Center of Value." In *Moral Principles in Action: Man's Ethical Imperative*, edited by Ruth Nanda Anshen. Science and Culture Series, vol. 5. New York: Harper & Bros, pp. 162–175.

Ostrom, E. (2007) "A diagnostic approach for going beyond panaceas." *Proceedings of the National Academy of Sciences*, 104(39): 15181–15187.

Ostrom, E. (2009) "A general framework for analyzing sustainability of social-ecological systems." *Science*, 325(5939): 419–422.

Otto, R. (1958) *The Idea of the Holy*. New York: Oxford University Press.

Peirce, C. (1960) *Collected Papers of Charles Sanders Peirce*. Harvard, MA: Belknap Press.

Pohl, C., Krütli, P., and Stauffacher, M., 2017 "Ten reflective steps for rendering research societally relevant." *GAIA*, 26: 43–51.

Popper, K. (1972/1979) *Objective Knowledge: An Evolutionary Approach*. Oxford: Oxford University Press.

Popper, K. (1982) *Realism and the Aim of Science*. London: Hutchinson.

Popper, K. (1990) *A World of Propensities: Two New Views of Causality*. Reading: Eastern Press.

Rawls, J.B. (1971) *A Theory of Justice*. Cambridge, MA: Harvard University Press.

Raworth, K. (2017) *Doughnut Economics: 7 Ways to Think Like a 21st Century Economist*. White River Junction, VT: Chelsea Green Publishing.

Raymond, C.M. and Kenter J.O. (2016) "Transcendental values and the valuation of ecosystem services." *Ecosystem Services*, 21: 241–257.

Rockström, J., Steffen, W., Noone, K., Persson, A.A., Chapin, F.S.*et al.* (2009) "A safe operating space for humanity." *Nature*, 461: 472–475.

Rolston, H., III (2007) "Living on Earth: Dialogue and Dialectic with My Critics." In *Nature, Value, Duty*, edited by C.J. Preston and W. Ouderkirk. Dordrecht: Springer, pp. 237–268.

Schumacher, E.F. (1973) *Small is Beautiful*. London: Blond & Briggs.

Schwab, K. (2016) *The Fourth Industrial Revolution*. New York: Crown Publishing.

Sellars, R.W. (1946) "Positivism and Materialism." *Philosophy and Phenomenological Research*, 7(1): 12–41.

Solow, R. (1974) "The Economics of Resources or the Resources of Economics." *American Economic Review*, 64(2): 1–14.

UN (1992) "United Nations Conference on Environment & Development Rio de Janeiro, Brazil, 3 to 14 June 1992: Agenda 21." https://sustainabledevelopment.un.org/content/documents/Agenda21.pdf.

UN (2012) *Review of Implementation of Agenda 21 and the Rio Principles: Synthesis*. New York: United Nations Department of Economic and Social Affairs.

UN (2019) *The Sustainable Development Goals Report*. New York: United Nations.

UN Environment (2019) *Global Environment Outlook – GEO-6: Healthy Planet, Healthy People*. https://wedocs.unep.org/20.500.11822/27539.

WCED (1987) *Our Common Future*. United Nations World Commission for Environment and Development. Oxford: Oxford University Press.

Whitehead, A.N. (1929) *The Function of Reason*. Boston, MA: Beacon Press.

Wilkinson, B.H. and McElroy, B.J. (2007) "The Impact of Humans on Continental Erosion and Sedimentation." *Geological Society of America Bulletin*, 119: 140–156.

Wilson, E.O. (2002) *The Future of Life*. New York: Alfred A. Knopf.

Wohlleben, P. (2017) *The Secret Wisdom of Nature: Trees, Animals, and the Extraordinary Balance of All Living Things Stories from Science and Observation.* Vancouver: David Suzuki Institute/Berkeley, CA: Greystone Books.

World Bank (2020) *Poverty and Shared Prosperity 2020: Reversals of Fortune.* Washington, DC: World Bank. https://doi.org/10.1596/978-1-4648-1602-4.

1 What is sustainability?

The "sustainability" concept has a long history dating back to ancient writers such as Plato, Strabo, Columella, and Pliny the Elder who, according to historian Du Pisani, "were not only aware of environmental degradation, but also recommended what we would call sustainable practices to maintain the 'everlasting youth' of the earth" (Du Pisani, 2006). From the beginning, sustainability implied that humans and the Earth's environment were in a dependence relationship, with the Earth providing life support and humans expected to respond in a way that would preserve the Earth's "everlasting youth".

In the eighteenth century, as population growth was seen to highlight natural resource scarcity, classical economists framed the sustainability idea in terms of equilibrium between the production process and the natural resources used as inputs. For instance, Thomas Malthus, in his 1798 *Essay on the principle of population as it affects the future improvement of society*, recommended checks on the exponential growth of population to protect the scarce agricultural land's capacity to provide food (Malthus, 1926). Concerns over the overuse of scarce natural resources threatening human welfare were also expressed in the nineteenth century by Stanley Jevons, G.P. Marsh, and J.S. Mill (Du Pisani, 2006). In 1848, in his book *Principles of political economy*, Mill recommended checks not only on population growth but also on capital stock growth in order to protect "the spontaneous activity of nature" and the "earth's pleasantness" (Mill, 1883). He envisioned an economy in a "stationary state", which was not a "stagnant" economy, but an economy in which humans would pursue "moral and social progress instead of material expansion" (El Serafy, 2013). Mill's "stationary" economy concept was a precursor to the "steady-state economy" concept later developed by ecological economist Herman Daly, and defined as "one in which population and the capital stock were not growing, even though the art of living continued to improve" (Daly, 1973, 2017; Daly and Cobb, 1994).

At the turn of the twentieth century, as industrialization in the USA was leading to impressive economic growth, the meaning of sustainability was impacted by a public debate concerning the proper use of natural resources. Should natural resources be "conserved" through "wise use", or should they be "preserved" or "locked up" from humans? The "conservation" side of the

DOI: 10.4324/9781003307587-2

debate was represented by Gifford Pinchot (1865–1946), one of the first American professional foresters, who advocated that natural resources, such as forests, be managed for sustainable use, meaning a utilitarian-type management that, according to Jeremy Bentham, would provide "the greatest good to the greatest number for the longest time" (Banzhaf, 2016). The "preservation" side, represented by environmentalist John Muir (1838–1914), advocated that natural resources be sheltered from use and kept in their wild state. The debate was won by the "conservationist" side, when a US Congress declared in 1896 that the purpose of the US forest reserves was "to furnish a continuous supply of timber" plus sustainable mining and grazing (Banzhaf, 2016). This anthropocentric and utilitarian understanding of sustainability as "conservation" of natural resources through "wise use" has remained the current practice in natural resource management and environmental policies of most countries. While the preservationist side lost the debate, John Muir's environmental activism raised awareness about the intrinsic value of renewable natural assets and led to the establishment of natural parks and wilderness areas starting with the USA and then spreading to other countries. It also inspired further "preservationist" thinking in other scientists' work, such as the ecologist Aldo Leopold and the economist John Krutilla. Aldo Leopold distrusted the ability of humans to "wisely use" natural resources when led only by economic utilitarian reasons. He believed that a new "ethic" was needed, one to regulate how "Humans could live with wildlife as part of a complex web of interactions" (Banzhaf, 2016). John Krutilla, an economist and himself a lover of wilderness, introduced new thinking in economics by insisting that overuse of natural resources had a real opportunity cost, expressed in the impossibility of "providing for the present and future the amenities associated with unspoiled natural environments" (Krutilla, 1967). He introduced the concepts of "existence" value and "option" value in economics and is considered the initiator of the new discipline of environmental and natural resource economics (Banzhaf, 2016).

In the 1960s, rapid technological and economic advances, based on massive consumption of natural resources, led to increased concerns about the consequences of fast economic growth for the Earth's sustainability, in books by Rachel Carson (1962), Kenneth Boulding (1966), Paul Ehrlich (1968), and Ernst Friedrich Schumacher (1973). In his book *Small is Beautiful: Economics as if People Mattered*, Schumacher wrote that humans "were very rapidly using up a certain kind of irreplaceable capital asset, namely the *tolerance margins* which benign nature always provides" (Schumacher, 1973). Early warnings about unlimited economic growth being unsustainable were expressed in 1971 by Simon Kuznets, the father of the Gross National Product, who described growth as economic expansion that is not sustained, whenever countries are "selling fortuitous gifts of nature to others" (El Serafy, 2013). The stringent need to limit unchecked economic growth was placed on the global political agenda in 1972 with the publication of the *Limits to Growth* report by a group of intellectuals organized as the Club of Rome. This report, based on a computerized model with 12 scenarios, predicted that overuse of non-renewable resources by an increasing world population, using a larger stock of human-made capital, mostly in modern industrial societies, will lead to ecological catastrophe and collapse of the current civilization within the next one hundred years, when the Earth's population was expected to reach and overshoot the carrying capacity of the Earth. The report defined sustainability as a "state of global equilibrium" with population and capital essentially stable, and "the forces tending to increase or decrease them in a carefully controlled balance" (Meadows et al., 1972). The report did not recommend a halt of economic growth; instead, it expressed the hope that "a rational and enduring state of equilibrium by planned measures" could be reached, or "could be designed so that the basic material needs of each person on earth are satisfied" (Meadows et al., 1972). However, the report noted that achieving a state of global equilibrium required "a basic change of values and goals at individual, national and world levels" (Meadows et al., 1972). A 20-year update (Meadows et al., 1992) and a 30-year update of the report (Meadows et al., 2005), confirmed the conclusions of the initial report, that there are demonstrated "general consequences of particular civilizational goals and strategies for achieving them, enacted within a finite world" (Floyd and Zubevich, 2010). In 2022, building on the systems-thinking legacy of the *Limits to Growth* report, a new report to the Club of Rome entitled *Earth for All: A Survival Guide for Humanity* was developed by a team of scientists and economists aiming to show that the collapse predicted by the 1972 report can be avoided if we "update" our economic thinking and we "redesign economic and social policies to put our societies on a pathway towards wellbeing for all within planetary boundaries" (Dixson-Declève et al., 2022). The five "turnarounds" proposed by the report in order to achieve these goals are: eliminate poverty, reduce inequality, empower women, transform food systems, and redesign energy systems. Missing from the "turnarounds" are the deep cultural and spiritual revolutions needed in our societies to become sustainable, as initially suggested in the 1972 *Limits to Growth* report which recommended "a basic change of values and goals at individual, national and world levels".

In the 1980s, the main contribution to the "sustainability" discussion came through ecological economics, a transdisciplinary field which "brought to the fore the idea of long-run 'limits to growth' owing to the inexorable substitution of free energy for bound energy in the production process" (Gale, 2018). The new field of ecological economics was born with the publication in 1971 of the seminal book *The Entropy Law and the Economic Process* by the mathematician-turned-economist Nicholas Georgescu-Roegen. The book challenged neoclassical economics' dogmatic reductionism which conveniently leaves out nature from its modelling and deals only with the "mechanics of utility and self-interest" (Georgescu-Roegen, 1975). Nicholas Georgescu-Roegen argued that both the biological and the economic processes are entropic in nature, being ruled by the second law of thermodynamics: "from the purely physical viewpoint, the economic process is entropic: it neither creates nor consumes matter or energy, but only transforms low into high entropy. But the whole physical process of the material environment is entropic too" (Georgescu-Roegen, 1971). The Georgescu-Roegen book identified a new absolute scarcity, the amount of low-entropy matter and energy existing in nature which is vital for any economic activity whose goals are for people "to stay alive and to keep a place under the social sun" (Georgescu-Roegen, 1971). This amount sets objective limits to economic activity and makes the idea of continuous economic growth within a non-growing ecosystem seem absurd: "low-entropy matter-energy is the physical coordinate of usefulness; the basic necessity that humans must use up but cannot create, and for which the human economy is totally dependent on nature's services" (Daly, 2007). Seeing the economy as a sub-system embedded in and dependent on the Earth's biosphere, ecological economics raised awareness of the need to preserve the aggregate environmental capital provided by nature because "we depend on our physical surroundings for all our activities, we stand literally empty handed if we lose vital functions" (Hueting and Reijnders, 1998). The Hueting and Reijnders article defined sustainability as "the use of vital environmental functions in such a way that they remain available indefinitely" and charged natural scientists with the task of scientifically establishing the "admissible" environmental burdens as objective standards for limiting economic activity (Hueting and Reijnders, 1998). Around the same time, another economist, El Serafy, argued that economists should calculate and explicitly include in countries' national accounts nature's "depreciation" or "the cost of safeguarding, repairing and maintaining these environmental functions" as a necessary addition to the national financial accounts, making them able to signal the value of the environmental functions provided by nature and the cost of their loss or diminution (El Serafy, 1998). While ecological economics did not recommend a cessation of economic growth, it argued for constraining "throughput" growth, where "throughput" is defined as the flow of matter and energy from environmental sources to nature's sinks, due to human economic activities (Daly, 1990). Limiting "throughput" involved establishing an optimal "scale" of macroeconomic activity, calibrated to not exceed the supporting ecosystem's limits (Daly, 2007). In 1991, another seminal ecological economics article defined sustainability as:

> a relationship between dynamic human economic systems and larger dynamic, but normally slower-changing ecological systems, in which (a) human life can continue indefinitely, (b) human individuals can flourish, and (c) human cultures can develop; but in which effects of human activities remain within bounds, so as not to destroy the diversity, complexity, and function of the ecological life support system.
>
> (Costanza et al., 1991)

In the 1990s, a series of studies have been performed on biodiversity and its impact on ecosystem functions and services (Schulze and Mooney, 1993, Tilman and Downing, 1994, Daily, 1997), clearly demonstrating that loss of certain life forms can impact both ecosystem functions and the services they provide to humanity. In 1997, a thought-provoking article estimated the average value of the world's ecosystem services at US\$33 trillion, much higher than the year's global Gross Domestic Product of US\$18 trillion (Costanza et al., 1997). A series of subsequent United Nations studies followed, aiming to understand and model natural capital and ecosystem services, such as the Millennium Eco-system Assessment (MEA) study (MEA, 2005) and the Economics of Ecosys-tems and Biodiversity (TEEB) report (TEEB Foundations, 2010). The idea that nature's wealth should be maintained on its own and be treated as the "limiting factor" of economic activity has been dubbed as "strong" sustainability (Huet-ing, 1974/1980; Goodland, 1995). "Strong" sustainability challenges the concept of "weak" sustainability favoured by mainstream economists. "Weak" sustain-ability aims to keep the "utility" (level of needs and wants) satisfaction non-declining over generations, by "maintaining a non-declining stock of economic capital into the indefinite future" (Norton, 2005), and by allowing "unlimited substitution" between natural and man-made types of capital (Solow, 1974) in the production of goods and services. The "weak" sustainability concept has been criticized for being "a social construct", reflecting "how the current and future quality of the environment is subjectively valued by an individual or group" (Hueting and Reijnders, 1998), and for being an "unworkable" concept as "utility is an experience, not a thing. It has no unit of measure and cannot be bequeathed from one generation to the next" (Daly, 2007). However, the "weak" sustainability approach has been incorporated into the "sustainable develop-ment" concept developed by the United Nations World Commission on Envir-onment and Development (WCED) in the 1987 report *Our Common Future* (WCED, 1987). The report famously defined sustainable development as "development that meets the needs of the present without compromising the ability of future generations to meet their own needs" (WCED, 1987).

The rise of sustainable development

The "sustainable development" concept, as developed by the WCED, borrowed from the 1972 *Limits to Growth* report and used similarly a global systemic approach with the same focus on development as the main path for reaching the fulfilment of humankind's needs. However, it fell short of defining environmental limits, which "are critical to sustainable development" (Meadowcroft, 2017).

The report acknowledged that sustainable development involves a process of planned change, but it envisioned change calibrated only in terms of ful-filment of growing needs, not in terms of existing natural resources. Accord-ing to the report, sustainable development is "a process of change in which the exploitation of resources, the direction of investments, the orientation of technological development, and institutional change are made consistent with future as well as present needs" (I.3.30). Thus, the "limits" identified in the report are not the natural limits of the environment but the "limitations imposed by the state of technology and social organization on the environment's ability to meet present and future needs" (WCED, 1987). While the report added a generous inter-generational equity dimension, it failed to address the most difficult issue of sustainable development, namely how to reconcile the objective limits of the resources available in the environment with the potentially unlimited needs and aspirations of humankind now and in the future. Reconciling the two "requires normative choice and political argument. It implies decisions about risk, the distribution of costs and bene-fits, and preferred ways of life" (Meadowcroft, 2017). The WCED report acknowledged the interdependence between the "environment" ("where we live") and development ("what we all do in attempting to improve our lot within that abode") and the fact that "the two are inseparable" (Brundtland, 1987). However, it made clear that what is to be sustained is not the envir-onment, but "the process of development itself" (Meadowcroft, 2017). The tool chosen to do so was economic growth, measured as crude growth in gross domestic product, backed by "policies that sustain and expand the environmental resource base. And we believe such growth to be absolutely essential to relieve the great poverty that is deepening in much of the devel-oping world" (Brundtland, 1987). The report's solution for attaining sustain-able development was a political one (Mebratu, 1998), consisting of a better integration of environmental policies with economic and social policies within the framework of sustainable development (WCED, 1987). Among the seven "critical objectives" for the integrated environment and development policies (reviving growth; changing the quality of growth; meeting essential needs for jobs, food, energy, water, and sanitation; ensuring a sustainable level of population; conserving and enhancing the resource base; reorienting technol-ogy and managing risk; and merging environment and economics in decision making) (WCED, 1987), only two were linked to the environment but in ambiguous terms, such as "conserving and enhancing the resource base". As for the change of values recommended by the *Limits to Growth* report, the WCED report referred only to the values underlying consumption, mention-ing that a "global rebalancing in consumption" (Gale, 2018) is needed, through the "promotion of values that encourage consumption standards that are within the bounds of the ecological possible and to which all can reason-ably aspire" (WCED, 1987).

The term "sustainable development" has become, since its introduction in 1987, an important and "essentially contested concept" (Gallie, 1955), given the numerous meanings attached to it. Some authors consider it "an ethical guiding principle and leading aspiration of humankind in the 21st century" (de Vries, 2012), while others consider it a tool for "improving the quality of human life while living within the carrying capacity of supporting ecosys-tems" (IUCN-UNEP-WWF, 1991). The concept has since been criticized for being "a normative abstraction" where, in the absence of objective criteria, "rules will be worked out over time via a competition of beliefs and moral debate" (Gladwin et al., 1995), and for being confusing since it is not clear to what it refers: "sustainable development of what: personal income, social complexity, gross national product, material frugality, individual consumption, ecological biodiversity?" (Luke, 1995). For others, sustainable development has become inefficient in making the sustainability transition happen, because "the world followed in words, but not in actions the WCED recommendations" (de Vries, 2019), as the concept was dubbed difficult to operationalize due to its strongly normative nature: "it is impossible to define sustainable development in an operational manner in the detail and with the level of control presumed in the logic of modernity" (Norgaard, 1994).

Even more socially damaging, the proliferation of diverse meanings attached to the "sustainable development" concept was considered a justification for unfair social practices, as "there are many constituencies which perceive the term 'sustainable development' as a vehicle to perpetuate many and varied corporate and institutional interests whilst giving the impression of adherence to, and observance of, environmentally-sound principles" (Johnston et al., 2007). In spite of these noted ambiguities of the concept, the term "sustainable development" with the three pillars of sustainability – environmental, social, and economic (Figure 1.1) – has been adopted since the 2005 UN World Summit as "the overarching framework of UN activity" (Farley and Smith, 2020).

In 2015, the 193 UN member states unanimously adopted the resolution *Transforming our world: the 2030 Agenda for Sustainable Development*. The resolution contains 17 aspirational goals with 169 targets for achieving, by 2030, social, economic, and environmental sustainable development worldwide, for the

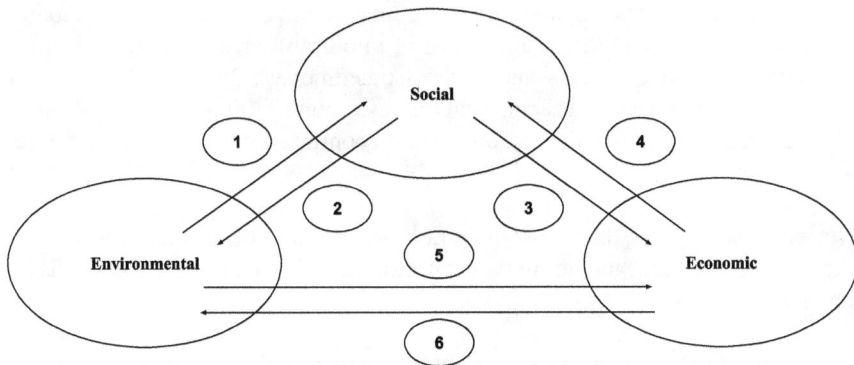

Figure 1.1 The three pillars of sustainability: environmental, social, and economic
Notes:
1. Dangers for the health of the population; impact on living and labour conditions.
2. Pressures on environmental resources; citizens' awareness with respect to environmental problems.
3. Quality and quantity of labour force; consumption.
4. Income distribution; number of jobs available.
5. The environment's involvement in production (resources and waste disposal).
6. Pressures on environmental resources; investments in environmental protection.
Source: Adapted from OECD, 1999.

people, planet, and prosperity. Implicit in the 17 goals is that we can ensure global prosperity and equality only within a stable ecological context, meaning that the integrity of the Earth's life-support systems must be maintained. However, critics note that the goals "fail to recognize that planetary, people and prosperity concerns are all part of one earth system, and that the protection of planetary integrity should not be a means to an end, but an end in itself" (Biermann et al., 2022). The goals continue to advocate for "sustained economic growth, higher levels of productivity and technological innovation" (UNDP, 2015) as the way to achieve global sustainability, a practical impossibility given the limited Earth ecosphere.

The literature on sustainability is growing. A 2007 article estimated that over 3,000 academic papers on sustainability were being published each year and the number of publications was increasing exponentially (Kajikawa et al., 2007). Over the last 20 years, several journals on sustainability have started publication, specializing in sub-domains related to sustainable agriculture, forestry, tourism, energy, and education (Kajikawa et al., 2007). Several books have also been written on sustainability, such as Kopnina and Shoreman-Ouimet, 2015; Fiscus and Fath, 2019; Ekardt, 2020; Farley and Smith, 2020. However, the definitions of "sustainability" are very diverse and demonstrate that "sustainability" itself remains an essentially contested concept, "the proper use of which inevitably involves endless disputes about [its] proper uses on the part of their users" (Gallie, 1955). Paul Ekins defines environmental sustainability as maintenance of natural capital:

> [T]he concept of sustainability itself simply means that whatever is being considered has the capacity for continuance. Environmental sustainability may be defined as the maintenance of important environmental functions and the natural systems and cycles (sometimes generically called natural capital) which generate them, such that their contributions to human health, wealth and wellbeing may be projected to continue into the indefinite future.
>
> (Ekins, 2017)

Herman Daly agrees that "sustainability means living on income and not on capital consumption, including consumption of natural capital" (Daly, 2017). For Kopnina and Shoreman-Ouimet:

> [S]ustainability means the capacity to support, maintain, or endure; it can indicate both a goal and a process. In ecology, sustainability describes how biological systems remain diverse, robust, resilient and productive over time, a necessary precondition for the well-being of humans and other species.
>
> (Kopnina and Shoreman-Ouimet, 2015)

For Ariane König, "the concept of sustainability is normative, as it suggests a direction in terms of 'good' or 'bad' ways in which society and environment interact" (König, 2018). Furthermore, a focus on the human–nature interface requires us to embrace complexity, in the diversity of perspectives and commitments (König, 2018). Another approach to sustainability focuses on sustainability as biotic interdependence in the web of life, where life is seen as "the 'thing' (quality, reality or process) we humans all share, it is also the 'thing' that unites the human species with all other species on Earth" (Fiscus and Fath, 2019), and which depends on participation of all living creatures, as "It is a law of the natural universe that no being can exist on its own resources. Everyone, everything, is hopelessly indebted to everyone and everything else" (Lewis, 2014). For Durning and for Beauregard, the definition of sustainability is very similar to the systemic definition of sustainable development: "In terms of sustainability, a moderate level of consumption, together with strong social institutions and a healthy environment, represents a better ideal than ever-increasing consumption" (Durning, 1992); and "sustainability is situated at the intersection of environmental protection, economic growth, and social justice" (Beauregard, 2003).

Where is sustainability?

The "sustainable development" concept and the "sustainability" concept continue to be used interchangeably, even if some authors (Mebratu, 1998; Robinson, 2004; Johnston et al., 2007) consider that the two concepts reflect very different realities and should not be used interchangeably. Ecological economics scholars have documented the impossibility of continued physical growth in a limited ecosystem (Daly and Farley, 2011; Higgs, 2014), as required by sustainable development. In terms of solutions, numerous authors advocate for degrowth (van den Bergh and Kallis, 2012), for development without growth (Victor, 2008; Jackson and Victor, 2020), or for a steady-state economy (Daly, 1991; Czech, 2006). Indeed, the wisdom of conflating "sustainable development" and "sustainability" and the narrative behind sustainable development are being increasingly questioned (Robèrt et al. 2002; Washington, 2015; Kajikawa et al., 2007; de Vries, 2019). Robèrt et al. (2002), for instance, have proposed a strategic approach to sustainable development, based on four sustainability principles rooted in the laws of thermodynamics and in studies of human behaviour. The principles, known as The Natural Step (TNS), state:

> In the sustainable society, nature is not subject to systematically increasing…
> 1. … concentrations of substances extracted from the Earth's crust
> 2. … concentrations of substances produced by society
> 3. … degradation by physical means, and
> 4. people are not subject to conditions that systematically undermine their capacity to meet their needs.
>
> Robèrt et al. (2002)

Also, de Vries considers that the modernity worldview behind the Sustainable Development Goals (SDGs) "no longer offers satisfactory principles and rules for the relationships of human beings with each other and with the natural world in the Anthropocene" (de Vries, 2019). He proposes "to enrich the one-sidedness of the SDGs and targets with values, beliefs, ethics and narratives from other worldviews" in a "rebalancing act in which the subjective and the immaterial are regaining legitimacy and the relation between individual and collective is reorientating itself" (de Vries, 2019).

Since the late 1990s, sustainability topics have been tackled by a distinct scientific field entitled "sustainability science" (Kates et al., 2001; Clark and Dickson, 2003; Komiyama and Takeuchi, 2006; de Vries, 2012; König, 2018). Sustainability science is an emerging field of inquiry seeking to understand "the fundamental character of interactions between nature and society" (Kates et al., 2001), and "a problem-driven and solution-oriented field that follows a transformational agenda" (Lang et al. 2012). This research field is important, as it promotes transdisciplinary research, which aims at "bridging the gap between problem solving and scientific innovation" (Lang et al. 2012) by bringing together insights from multiple disciplines and from non-academic stakeholders or knowledge users. However, sustainability science faces the challenging task of bridging the gap between the objective, value-neutral character of natural sciences describing nature, and the world views and values that are integral to the concept of sustainability, where normative questions like "what is to be sustained, and at what scale, and in what form" (Lélé and Norgaard, 1996) are unavoidable. Some themes approached in the sustainability science literature are transdisciplinary research (Lang et al., 2012; Brandt et al., 2013); sustainability science as a transformative social learning process (König, 2018); sustainability assessment (Pinter et al., 2012, Sala et al., 2015); and values and their role in transformational sustainability science (Horcea-Milcu et al., 2019).

Sustainability is not sustainable development

I hope it has by now become clear that sustainability is not sustainable development. The two concepts are fundamentally different, and it is sustainability and not sustainable development that should be the overarching goal of ecological, social, economic, and cultural policies on a thriving planet with prosperous successive generations. The concept of "food security" indicates "sustainability" when the UN Food and Agriculture Organization (FAO) calculates the minimum caloric needs of 2,100 kilocalories per person, per day and the recommended requirements of protein, fat, and carbohydrates. It indicates "sustainable development" when it is "shaped by the political and economic structures within which FAO is situated" (Iversen et al., 2023).

In my view, the main differences between the two concepts are as follows. Sustainable development is a subjective social construct about how humans choose to use nature, including the Earth's life-sustaining system. The ambiguity in defining the concept can lead to poor or deceptive choices by many actors. The "needs" of current and future generations are not clearly defined and are subject to interpretation, as their definition implies "an inextricable combination of value judgments, worldviews, and consensual knowledge" (Lélé and Norgaard, 1996). The three-pillar sustainable development structure is arbitrary in the absence of clearly defined ecological limits, as it can imply "more growth, more nature conservation, more kindergartens" (Ekardt, 2020).

Economic growth and technological innovation are considered the engines of sustainable development, in spite of the ecological destruction and increase in social inequality produced by unlimited economic growth on our finite planet over the past 35 years. Many people believe that ecosystems are sufficiently resilient to adapt, in time, to the changes we have inflicted upon them. This is "a perverse depiction of a sustainable future world but one to which some nonetheless cling tenaciously" (Johnston et al., 2007). The sustainable development concept ignores the necessary change in values which has been recommended by the Limits to Growth report since 1972. To sum up, I agree with Kajikawa et al. that "While sustainable development is associated with the human exploitation of nature, 'sustainability' does not include such a connotation" (Kajikawa et al., 2007).

Sustainability is not only a theoretical concept about humans "living in harmony with nature and with one another" (Mebratu, 1998), but also an objective feature of the world, a numinous condition that makes life on planet Earth possible and meaningful for this and future human generations:

> The biospheric membrane that covers the Earth, and you and me, is the miracle we have been given. And our tragedy, because a large part of it is being lost forever before we learn what it is and the best means by which it be savored and used.
>
> (Wilson, 2002)

Sustainability is embodied in structures, functions, and processes specifically crafted to protect and sustain life. Some are visible, such as birds migrating to warmer climates before winter, guided by the Earth's magnetic inclination (Wynn et al., 2022); some are invisible, such as photosynthesis, metabolism, or the immune system of living beings. The "ecological capital of the Earth" can be scientifically established, as we know from Einstein that "the eternal mystery of the world is its comprehensibility [...] The fact that it is comprehensible is a miracle." (Einstein, 1936). It is important that natural scientists establish the maximum environmental burden as fixed reference points for human use and development (Hueting and Reijnders, 1998). Not only natural laws, such as the law of gravity, or the law of entropy (Georgescu-Roegen, 1971), should be treated as boundary conditions, but also data provided by human measurements such as the ecological footprint (Wackernagel and Rees, 1995), the planetary boundaries (Rockström et al., 2009; Steffen et al., 2015), or the global tipping points (Lenton, 2020). Sustainability implies that life in all forms is precious, it is worth sustaining, and that humans, as the only living things on Earth endowed with reason and "the moral law within" (Kant, 1788/1997), are expected to participate responsibly in the life-web at the core of sustainability, obeying the rules of functioning of the physical Universe. One of the first rules is that "planetary integrity is a global public good that is maintained by keeping the earth system within its ecological limits" (Westra, Bosselmann and Gwiazdon, 2018). While science can help establish these ecological limits and discover new ones in the quantum field, for instance, it cannot explain sustainability's "numinous" condition that biologist Edward O. Wilson referred to, unless we are willing to enlarge the field of science, defining it, as Einstein did:

> The whole of science is nothing more than a refinement of every day thinking. It is for this reason that the critical thinking of the physicist cannot possibly be restricted to the examination of the concepts of his own specific field.

(Einstein, 1936)

The working definition of sustainability that is proposed in this book is:

> *Sustainability is the fundamental nature of our Universe that continues to exist and function as an interconnected, complex, emergent and ontically open life-support system available to all living things. This includes rational, sentient, and moral human beings, who are free to choose the nature and level of involvement in the mysterious web of life, using it destructively or in harmony with other living creatures.*

We will consider adding philosophy, cosmology, ethics, and metaphysics to pure scientific explanations of sustainability, in an attempt to understand what sustainability really is: a gift giving meaning to perpetuation of life in many forms on planet Earth, a gift that *must* be explored and known by humans in order to properly steward it. Accepted as a gift, sustainability will enable humans to regain the awe that has been lost when they decided to ignore it and act as masters of nature and of their own destiny.

This chapter has explored the historical evolution of the "sustainability" concept, the emergence of the "sustainable development" concept, and the current confusion concerning the definitions and use of the two. It has explained how the two concepts are different, mainly because sustainability is not only a concept but an objective feature of our world, requiring both scientific and metaphysical approaches to be understood. This is important, as I argue that sustainability, not sustainable development, should guide human choices in dealing with ecology, social, economic, and cultural decisions and actions.

References

Banzhaf, H.S. (2016) "The Environmental Turn in Natural Resource Economics John Krutilla and "Conservation Reconsidered." Resources for the Future Discussion Paper, RFF 16-27: 1–23.

Beauregard, R.A. (2003) "Democracy, storytelling, and the sustainable city." In *Story and Sustainability*, edited by B. Eckstein and J.A. Throgmorton. Cambridge, MA: MIT Press.

Biermann, F., Hickmann, T., Sénit, CA.*et al.* (2022) "Scientific evidence on the political impact of the Sustainable Development Goals." *Nature Sustainability*, 5: 795–800. https://doi.org/10.1038/s41893-022-00909-5.

Boulding, K.E. (1966) "The economics of the coming spaceship earth." In *Environmental Quality in a Growing Economy*, edited by H. Jarret. Baltimore: Johns Hopkins University Press.

Brandt, P., Ernst, A., Gralla, F., Luederitz, C., Lang, D.J., Newig, J., Reinert, F., Abson, D.J., and von Wehrden, H. (2013) "A review of transdisciplinary research in sustainability science." *Ecological Economics*, 92: 1–15.

Brundtland, G.H. (1987) *Our Common Future, Chairman's Foreword*, https://sustainabledevelopment.un.org/content/documents/5987our-common-future.pdf.

Carson, R. (1962) *Silent Spring*. Boston: Houghton Mifflin.

Clark, W.C. and Dickson, N.M. (2003) "Science and Technology for Sustainable Development" special feature: "Sustainability science: the emerging research program." *Proceedings of the National Academy of Sciences*, 100(14): 8059–8061.

Connelly, S. (2007) "Mapping Sustainable Development as a Contested Concept." *Local Environment*, 12(3): 259–278.

Costanza, R., Daly, H.E., and Bartholomew, J.A. (1991) "Goals, agenda and policy recommendations for ecological economics." In *Ecological Economics: The Science and Management of sustainability*, edited by R. Costanza. New York: Columbia University Press, pp. 1–20.

Costanza, R., d'Arge, R., de Groot, R., Farber, S., Grasso, M., Hannon, B., Limburg, K., Naeem, S., Oneill, R.V., Paruelo, J., Raskin, R.G., Sutton, P., and van den Belt, M. (1997) "The value of the world's ecosystem services and natural capital." *Nature*, 387(6630): 253–260.

Czech, B. (2006) "Steady State Economy." In *Encyclopedia of Earth*, edited by T. Tietenberg et al. Washington, DC: National Council for Science and the Environment.

Daily, G.C. (1997) *Nature's Services: Societal Dependence on Natural Ecosystems*. Washington: Island Press.

Daly, H.E. (1973) *Toward a Steady-State Economy*. San Francisco: W.H. Freeman.

Daly, H.E. (1990) "Towards some operational principles of sustainable development." *Ecological Economics*, 2: 1–6.

Daly, H.E. (1991) *Steady-State Economics*. Washington, DC: Island Press.

Daly, H.E. (2007) *Ecological Economics and Sustainable Development Selected Essays of Herman Daly*. Cheltenham, UK and Northampton, MA: Edward Elgar.

Daly, H. (2017) "A new economics for our full world." In *Handbook on Growth and Sustainability*, edited by Brett B. Victor and B. Dolter. Cheltenham, UK: Edward Elgar.

Daly, H. and Cobb, J. (1994) *For the Common Good: Redirecting the Economy Toward Community, the Environment, and a Sustainable Future*, 2nd ed. Boston: Beacon Press.

Daly, H. and Farley, J. (2011) *Ecological Economics: Principles and Applications*. Washington, DC: Island Press.

de Vries, B.J.M. (2012) *Sustainability Science*. Cambridge: Cambridge University Press.

de Vries, B.J.M. (2019) "Engaging with the Sustainable Development Goals by Going Beyond Modernity: An Ethical Evaluation Within a Worldview Framework." *Global Sustainability*, 2(e18): 1–14.

Dixson-Declève, S., Gaffney, O., Ghosh, J., Randers, J., Rockström, J., and Stoknes, P.E. (2022) *Earth for All: A Survival Guide for Humanity. A report to the Club of Rome.* New Society Publishers.

Du Pisani, J.A. (2006) "Sustainable development – historical roots of the concept." *Environmental Sciences*, 3(2): 83–96.

Durning, A. (1992) *How Much is Enough? The Consumer Society and the Future of Earth.* New York: W.W. Norton and Co.

Ehrlich, P.R. (1968) *The Population Bomb.* New York: Ballantine Books.

Einstein, A. (1936) "Physics and Reality." *Journal of the Franklin Institute*, 22(3): 349–382.

Ekardt, F. (2020) *Sustainability Transformation, Governance, Ethics, Law.* Cham, Switzerland: Springer.

Ekins, P. (2017) "Ecological modernization and green growth: prospects and potential." In *Handbook on Growth and Sustainability*, edited by P.A. Victor and B. Dolter. Cheltenham, UK and Northampton, MA: Edward Elgar.

El Serafy, S. (2013) *Macroeconomics and the Environment Essays on Green Accounting.* Cheltenham, UK and Northampton, MA: Edward Elgar.

Farley, H.M. and Smith, Z.A. (2020) *Sustainability: if it's everything, is it nothing?* 2nd ed. New York: Routledge.

Fiscus, D.A. and Fath, B.D. (2019) *Foundations for Sustainability: A Coherent Framework of Life-Environment Relations.* London: Academic Press/Elsevier.

Floyd, J. and Zubevich, K. (2010) "Linking foresight and sustainability: An integral approach." *Futures*, 42: 59–68.

Gale, F.P. (2018) *The Political Economy of Sustainability.* Cheltenham, UK and Northampton, MA: Edward Elgar.

Gallie, W. (1955) "Essentially contested concepts." *Proceedings of the Aristotelian Society*, 56: 167–198.

Georgescu-Roegen, N. (1971) *The Entropy Law and the Economic Process.* Cambridge, MA: Harvard University Press.

Georgescu-Roegen, N. (1975) "Energy and Economic Myths." *Southern Economic Journal*, 41(3): 347–381.

Gladwin, T.N., Kennelly, J.J., and Krause, T.-S. (1995) "Shifting Paradigms for Sustainable Development: Implications for Management Theory and Research." *The Academy of Management Review*, 20(4): 874–907.

Goodland, R. (1995) "The concept of environmental sustainability." *Annual Review of Ecology, Evolution, and Systematics*, 26: 1–24.

Higgs, K. (2014) *Collision course: Endless growth on a finite planet.* Cambridge, MA: MIT Press.

Horcea-Milcu, A.-I., Abson, D.J., Apetrei, C.I., Duse, I.A., Freeth, R., Riechers, M., Lam, D.P.M., Dorninger, C., and Lang, D.J. (2019) "Values in transformational sustainability science: Four perspectives for change." *Sustainability Science*, 14, 1425–1437. https://doi.org/10.1007/s11625-019-00656-1.

Hueting, R., 1980 [1974]. *New Scarcity and Economic Growth.* Amsterdam, New York, Oxford: North Holland Publishing Company. (Original work published 1974)

Hueting, R. and Reijnders, L. (1998) "Sustainability as an Objective Concept." *Ecological Economics*, 27(2): 139–147.

IUCN-UNEPP-WWF (1991) *Caring for the Earth: A strategy for sustainable living.* IUCN-UNEP-WWF Report.

Iversen, T.O., Westengen, O.T., and Jerven, M. (2023) "The history of hunger: Counting calories to make global food security legible." *World Development Perspectives*, 30, 100504.

Jackson, T. and Victor, P. (2020) "The Transition to a Sustainable Prosperity-A Stock-Flow-Consistent Ecological Macroeconomic Model for Canada." *Ecological Economics*, 177.

Jacobs, M. (1999) "Sustainable development as a contested concept." In *Fairness and Futurity: Essays on Environmental Sustainability and Social Justice*, edited by A. Dobson. Oxford: Oxford University Press, pp. 21–45.

Johnston, P., Everard, M., Santillo, D., and Robèrt, K-H, (2007) "Reclaiming the Definition of Sustainability." *Environmental Science and Pollution Research*, 14(1): 60–66.

Kajikawa, Y., Ohno, J., Takeda, Y., Matsushima, K., and Komiyama, H. (2007) "Creating an academic landscape of sustainability science: an analysis of the citation network." *Sustainability Science*, 2: 221–231.

Kant, I. (1997) "Critique of Practical Reason." In *Cambridge Texts in the History of Philosophy*. Cambridge: Cambridge University Press. (Original work published 1788)

Kates, R.W., Clark, W.C., Corell, R., Hall, J.M., Jaeger, C.C., Lowe, I., McCarthy, J. J., Schellnhuber, H.J., Bolin, B., Dickson, M., Faucheux, S., Gallopin, G.C., Grübler, A., Huntley, B., Jäger, J., Jodha, N.S., Kasperson, R.E., Mabogunje, A., Matson, P., Mooney, H., Moore III, B., O'Riordan T., and Svedin, U. (2001) "Sustainability Science." *Science*, 292(5517): 641–642.

Komiyama, H. and Takeuchi, K. (2006) "Sustainability science: building a new discipline." *Sustainability Science*, 1: 1–6.

Kopnina, H. and Shoreman-Ouimet, E. (2015) "Introduction. The emergence and development of sustainability." In *Sustainability: Key issues*, edited by H. Kopnina and E. Shoreman-Ouimet. London and New York: Routledge, pp. 3–24.

König, A. (ed.) (2018) *Sustainability Science Key Issues*. London and New York: Routledge.

Krutilla, J. (1967) "Conservation Reconsidered." *American Economic Review*, 57(4): 777–786.

Lang, D.J., Wiek, A., Bergmann, M., Stauffacher, M., Martens, P., Moll, P., Swilling, M., and Thomas, C.J. (2012) "Transdisciplinary research in sustainability science: practice, principles, and challenges." *Sustainability Science*, 7(Supplement 1): 25–43.

Lélé, S. and Norgaard, R.B. (1996) "Sustainability and the Scientist's Burden." *Conservation Biology*, 10(2): 354–365.

Lenton, T.M. (2020) "Tipping positive change." *Philosophical Transactions of the Royal Society B (Biological Sciences)*, 375(1794): 1–8.

Lewis, C.S. (2014) *God in the Dock*. New York: HarperOne.

Luke, T. (1995) "Sustainable Development as a Power/Knowledge System: The Problem of 'Governmentality.'" In *Greening Environmental Policy*, edited by F. Fischer and M. Black. New York: Palgrave Macmillan. https://doi.org/10.1007/978-1-137-08357-9_2.

Malthus, T.R. (1926) *First essay on population*. London: Macmillan.

MEA (2005) *Millennium Ecosystem Assessment (MEA). Ecosystems and Human Well-Being: Synthesis*. Washington, DC: Island Press.

Meadowcroft, J. (2017) "Sustainable development, limits and growth: Reflections on the conundrum." In *Handbook on Growth and Sustainability*, edited by P.A. Victor and B. Dolter. Cheltenham, UK: Edward Elgar.

Meadows, D.H., Meadows, D.L., Randers, J., and Behrens, W.W. III (1972) *Limits to Growth*. Falls Church, VA: Potomac Associates/Universe Books.

Meadows, D.H., Meadows, D.L., and Randers, J. (1992) *Beyond the Limits – Global Collapse or a Sustainable Future.* London and Stirling, VA: Earthscan.

Meadows, D., Randers, J., and Meadows, D. (2005) *Limits To Growth: The 30-Year Update.* London and Stirling, VA: Earthscan.

Mebratu, D. (1998) "Sustainability and sustainable development: historical and conceptual review." *Environmental Impact Assessment Review,* 18, 493–520.

Mill, J.S. (1883) *Principles of political economy, with some of their applications to social philosophy.* London: Longmans, Green & Co.

Norgaard, R.B. (1994) *Development Betrayed.* London: Routledge.

Norton, B.G. (2005) *Sustainability. A philosophy of adaptive ecosystem management.* Chicago, IL and London: University of Chicago Press.

OECD (1999) *The OECD Three-Year Project on Sustainable Development: A Progress Report.* Paris: OECD.

Pinter, L., Hardi, P., Martinuzzi, A., and Hall, J. (2012) "Bellagio STAMP: principles for sustainability assessment and measurement." *Ecological Indicators,* 17: 20–28.

Robèrt, K-H, Schmidt-Bleek, B., Aloisi de Larderel, J., Basile, G., Jansen, J.L., Kuehr, P., Thomas, P., Suzuki, M., Hawken, P., and Wackernagel, M. (2002) "Strategic sustainable development – Selection, design and synergies of applied tools." *Journal of Cleaner Production,* 10(3): 197–214.

Robinson, J. (2004) "Squaring the circle? Some thoughts on the idea of sustainable development." *Ecological Economics,* 48: 369–384.

Rockström, J., Steffen, W., Noone, K.*et al.* (2009) "A safe operating space for humanity." *Nature,* 461: 472–475. https://doi.org/10.1038/461472a.

Sala, S., Ciuffo, B., and Nijkamp, P. (2015) "A systemic framework for sustainability assessment." *Ecological Economics,* 119: 314–325.

Schulze, E.D. and Mooney, H.A. (1993) *Biodiversity and Ecosystem Function.* Berlin: Springer.

Schumacher, E.F. (1973) *Small is Beautiful.* London: Blond & Briggs.

Solow, R. (1974) "The Economics of Resources or the Resources of Economics." *American Economic Review,* 64(2): 1–14.

Steffen, W., Richardson, K., Rockström, J., Cornell, S.E., Fetzer, I., Bennett, E.M., Biggs, R., Carpenter, S.R., de Vries, W., de Wit, C.A., Folke, C., Gerten, D., Heinke, J., Mace, G. M., Persson, L.M., Ramanathan, V., Reyers, B., and Srlin, S. (2015) "Planetary boundaries: Guiding human development on a changing planet." *Science,* 347(6223): 1–10.

TEEB Foundations (2010) *The Economics of Ecosystems and Biodiversity: Ecological and Economic Foundations.* London and Washington: Earthscan.

Tilman, D. and Downing, J.A. (1994) "Biodiversity and stability in grasslands." *Nature,* 367: 363–365.

UNDP (2015) "What are the Sustainable Development Goals?" United Nations Development Programme. www.undp.org/sustainable-development-goals.

van den Bergh, J.C.J.M. and Kallis, G. (2012) "Growth, A-Growth or Degrowth to Stay within Planetary Boundaries?" *Journal of Economic Issues,* 46(4): 909–920.

Victor, P.A. (2008) *Managing Without Growth: Slower by Design, not Disaster.* Cheltenham, UK and Northampton, MA: Edward Elgar.

Wackernagel, M. and Rees, W.E. (1995) *Our Ecological Footprint: Reducing Human Impact on the Earth.* New Society Publishers.

Washington, H. (2015) "Is 'sustainability' the same as 'sustainable development'?" In *Sustainability Key Issues,* edited by H. Kopnina and E. Shoreman-Ouimet. London and New York: Routledge.

WCED (1987) *Our Common Future*. United Nations World Commission for Environment and Development. Oxford: Oxford University Press.

Westra, L., Bosselmann, K., Gray, J., and Gwiazdon, K. (eds) (2018) *Ecological Integrity, Law and Governance*. Abingdon and New York: Routledge.

Wilson, E.O. (2002) *The Future of Life*. New York: Alfred A. Knopf.

Wynn, J. *et al.* (2022) "Magnetic stop signs signal a European songbird's arrival at the breeding site after migration." *Science*, 375: 446–449.

2 Searching for the truth
Research methods

This book argues that sustainability is not only a theoretical concept about how the world functions, but also an objective feature of the world allowing humans to exist and societies to flourish. While sustainability in nature is easier to perceive and assess than in society, in both realms sustainability appears both in material and in immaterial forms. That is why the investigation methods should be able to correctly identify both the material dimension (facts discoverable through empirical investigation) and the immaterial dimension of sustainability, such as photosynthesis in biochemistry, or values, norms, motives, relationships, rules, and standards in society, which, though invisible, drive choices, behaviours, and policy decisions towards either sustainability or non-sustainability. In addition, sustainability points to some basic questions about ultimate reality which require a wider ontology and broader research methods beyond empirical research and quantitative measurements. These questions refer to the nature of nature, to life and its origin, to humans' place in nature and how much they can impact sustainability's mechanisms, as well as what we can scientifically know about sustainability, or what changes might be needed in humans' actions to solve our current issues related to unsustainable living. Should these changes take place at the level of the individual or at the level of social structures? Scientific materialism, based in modern positivism and epistemological empiricism, gives simplistic answers to these questions by affirming that only the natural world is out there, and it consists solely of matter and energy, and that the laws of nature can explain all things happening in the Universe (Carroll, 2017). However, discoveries in natural science (physics, chemistry, biology) have shown, starting from the discovery of quantum indeterminacy in physics (1927), that more than simple and effect is at work in the complex aggregate of nature that maintains the life-sustaining system. Then a challenging question arises, how can we reconcile in a coherent way "our overall *world view*, which includes our general notions concerning the nature of reality, along with those concerning the total order of the Universe, i.e., *cosmology*?" (Bohm, 1980). Bohm recommends that "our notions of cosmology and of the general nature of reality must have room in them to permit a consistent account of consciousness" (Bohm, 1980).

DOI: 10.4324/9781003307587-3

This is not how the modern science works, where reality is studied in disciplinary fragments and wholeness of reality is sacrificed for the sake of pragmatic outcomes and non-material entities are completely ignored. In social sciences, dominated by postmodern thinking and the widespread beliefs in relativism, subjectivity of facts, as well as in humans' ability to construct social reality (Berger and Luckmann, 1966; Creswell and Miller (2000), sustainability research has focused mainly on "ways to administer sustainable development within a society" (Farley and Smith, 2020), and a call to discern sustainability values through participatory research (Ratner, 2004; Gale, 2018). While this attempt at values-clarification is valuable, it comes with some dangers, represented by constructivist-interpretivist approaches of qualitative research "to allow sustainability to mean everything to everyone" (Farley and Smith, 2020) in the absence of some objective values to guide the change in behaviour required by sustainability.

These reasons warrant the adoption of a critical realist approach in our research, able both to describe in realistic terms the objectively existing sustainability in the world, and help us explain causalities or "observable regularities by appealing to underlying and often unobservable structures" (O'Hare, 1985). Such a research approach will allow us to cross the methodological borders between natural and social sciences, a requirement often expressed in current social research (Danermark et al., 2019), mainly in transdisciplinary research on sustainability. The critical realist approach is inspired by Karl Popper's realist epistemology, especially his theory of propensities or objective probabilities (Popper, 1990), and by Roy Bhaskar's critical realism (1975, 1979, 2010), a philosophical framework that rejects the epistemic fallacy (i.e. confusing the map for the territory), by promoting a clear distinction between ontology (the philosophical study of being, or *what is*) and epistemology (the philosophical study of *knowledge about being*). Critical realism is best fitted to explain sustainability by connecting the ontological realism of natural sciences with the epistemological relativism and judgmental rationality of social sciences, while also taking into account the possibility of human action through "a critical engagement with the 'cultural turn' and the importance of discourse to human action and identity and action" (Bhaskar, 2010). We need a critical realist approach in our research on sustainability because critical realism insists that there is an objective, "real" world that exists, one which is independent of human consciousness and not reducible to simply our knowledge of it. Reality in critical realism is "a state of the matter which is what it is, regardless of how we do view it, choose to view it or are somehow manipulated into viewing it" (Archer, 2007), while at the same time recognizing that epistemologically this world cannot always be known objectively because what we see is rationally processed through our brains, language, culture, methods, and so on (Westhorp, 2013).

Karl Popper's critical rationalism follows the tradition inherited from the Greek philosophers to freely and critically discuss various theories based in real-world problems as well as their proposed solutions, but expecting not to verify their truthfulness but to falsify them, as, according to Popper, in science we "cannot know, we can only guess. And our guesses are guided by the unscientific, the metaphysical (though biologically explainable) faith in laws, in regularities which we can uncover" (Popper, 1935/1959). Popper recommended as scientific method a process of "conjectures and refutations" consisting in developing hypotheses and testing them against competing hypotheses. Like Popper, I believe that "Next to music and art, science is the greatest, most beautiful and most enlightening achievement of the human spirit" (Popper, 1990), and I broadly define science as purposeful human activity to produce knowledge through a process of "discovery of complexity inherent in the natural and social reality" (Sabau, 2001). This implies that both natural and social sciences should adapt their research methods in the attempt to present realistic understandings of the complex reality that exists, and to understand the "big picture" of how the world functions, by crossing disciplinary boundaries that we humans continually create epistemologically. This has already happened, especially in research about socio-ecological systems such as forests, fisheries, or agricultural systems, where interdisciplinarity is accomplished through grounded theory, triangulation, or various systemic approaches (Nuijten, 2011). In developing scientific theories, scientists should be guided by an honest desire to learn, by the joy of discovery, by a willingness to follow the evidence where it logically leads, and to extend their sources of knowledge beyond the dry facts of positive science. Popper was right not to exclude myths and metaphysical ideas from his sources of knowledge, as he believed that "purely metaphysical ideas" have contributed to the advancement of science:

> From Thales to Einstein, from ancient atomism to Descartes's speculations about matter, from the speculations of Gilbert and Newton and Leibniz and Boscoviz about forces to those of Faraday and Einstein about field of forces, metaphysical ideas have shown the way.
>
> (Popper 1935/1959)

Popper believed that the scientists' effort to develop a scientific theory and to carefully design experiments and observations aimed at falsifying the formulated theory, as well as the scientists' earnest desire to participate in the public discussion of their theories are essential not only for the growth of knowledge but also for achieving the scientific objectivity of the theories, and in the end for fulfilling the main role of science which is "the search for truth". Popper was aware that in science, especially in social sciences, the truth of theories is difficult to prove, as "the results of science remain hypotheses that may have been well-tested, but not established: not shown to be true. Of course, they may be true. But even if they fail to be true, they are splendid hypotheses, opening the way to still better ones" (Popper, 1990). However, he believed that this difficulty is not a valid reason for not attempting to arrive closer to the truth in scientific research. Popper saw the value of interdisciplinary research and recommended that natural scientists and philosophers work together to paint a realistic picture of the world and to solve real-world problems. Acknowledging the difficulty of crossing disciplinary borders, Popper advised scientists to be open-minded and generous, admitting that "I may be wrong and you may be right, and by an effort, we may get nearer to the truth" (Popper, 1945), truth that he defined as "true descriptions of certain facts, or aspects of reality" (Popper, 1972).

Popper's concern with realistic explanations of our mind-independent external world, led him to his theory of propensities. In 1990, he wrote: "we live in a world of propensities, and [...] this fact makes our world both more interesting and more homely than the world as seen by earlier states of the sciences" (Popper, 1990). Popper formulated his theory of propensities by attempting to explain one of the most remarkable characteristics of our Universe, "the tendency of statistical averages to remain stable if the conditions remain stable". He explained that in the world of probabilities, there are "weighted possibilities which are more than mere possibilities, but tendencies or propensities to become real". For him, propensities were not only tendencies but "physical realities, even if invisible, which can act as Newton's attractive forces, or as fields of forces able to set bodies in motion and to produce change". He further explained that "propensities should not be regarded as properties inherent in an object, such as a die or a penny, but that they should be regarded as inherent in a situation (of which, of course, the object was a part)" (Popper, 1990). These situations can be identified in physics, in chemistry, biochemistry and in biology. For instance, Popper's discovery of propensities was important for a realist interpretation of the quantum theory in mechanics, in which Einstein's curved spacetime can be considered an objective propensity. But we should expect a world of indeterminacies and causal agents to also exist in social life, where even if humans can act as intentional self-conscious agents, "causation derives from the powers of structures, whether natural or social" (Gorski, 2013), and not only from human agents. We can use Popper's theory of propensities in our analysis of sustainability as a normative aim which potentially can be realized in a world of propensities in which

> the future is open: objectively open. Only the past is fixed; it has been actualized and so it has gone. The present can be described as the continuing process of the actualization of propensities; or, more metaphorically, of the freezing or the crystallization of propensities. While the propensities actualize or realize themselves, they are continuing processes.
> (Popper, 1990)

Very similar to Popper's world of propensities, in 1975, Roy Bhaskar described the world as "an open system consisting of things possessing causal powers or potentialities and liabilities in virtue of their intrinsic structures (essential natures)" (Bhaskar, 1975). These structures are real entities, such as the universal natural laws, which are not products of humans, but which can be discovered and investigated by science. Bhaskar's realism, called transcendental realism, is different both from David Hume's classical empiricism and Immanuel Kant's transcendental idealism. According to transcendental realism, "both knowledge and the world are structured, both are differentiated and changing" (Bhaskar, 1975). The world is stratified ontologically in three distinctive realms, the "real", the "actual", and the "empirical" (Figure 2.1), with the real encompassing the actual and the empirical domains, but also including various powers (natural mechanisms) and their potentialities, such as the specific gravity of mercury, the process of electrolysis, and the mechanism of light propagation (Bhaskar, 1975).

We can exemplify the "real" as material ecosystems, geological systems, whole cosmos systems that exist and endure independently of humans and their actions, and include the realm of the actual reality where the universal laws of nature act to produce events, and the realm of the empirical reality where humans observe and experience these events. For instance, we can experience a hurricane and observe all the visible phenomena related to it, such as high winds, clouds, heavy rain, and thunder and lightning (empirical domain), and be able to observe and measure some of the meteorological phenomena involved in the hurricane. We can measure the temperature, amount of fallen rain, and the wind speed, but be surprised by the amplitude of the actual storm surge, or flood (actual domain). Still, all this information may not be enough to explain the full causal mechanisms, due to the emergent properties, such as the "butterfly effect" (Lorenz, 1963), that have produced the hurricane in the middle of the Atlantic Ocean (real domain). The role of science is to produce not only empirical knowledge of the seen and experienced phenomena but also knowledge of the unseen powers, and mechanisms and their tendencies to produce phenomena in nature. These powers, tendencies, and generative mechanisms are called by Bhaskar (1975) "the structured and intransitive objects of scientific inquiry". In contrast to the intransitive objects of knowledge, the "transitive" objects of knowledge consist of prior existing knowledge which is historically, socially, and culturally informed, and is used by the researchers to produce new knowledge.

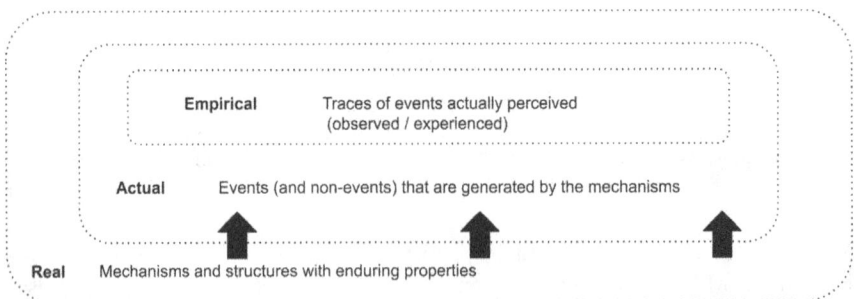

Figure 2.1 The stratified ontology of critical realism
Source: Smith and Johnston, 2014.

The transitive dimension of inquiry explains both the epistemic relativism and judgmental rationality in social science research, as the ways researchers come to discover the intransitive entities are "context-dependent, fallible and prone to individual bias and the vicissitudes of accidental properties of class, ethnicity, gender, sexuality" (Quraishi et al., 2022). At the same time, the researchers must make judgements and decisions about competing epistemic accounts of reality, without always having clear evaluative criteria and using their rational powers, which most of the time are subjective.

Bhaskar started his theory of critical realism by studying knowledge production in natural sciences, but he was also interested in social sciences and their inability to explain social phenomena and consequently solve social problems. He was a strong critic of *naturalism*, the theory that there is an essential unity of research method between the natural and social sciences, in both of its two forms, as *reductionism*, the belief that social beings' actions can be explained by physical laws, and as *scientism*, the theory that "insists on the need not only for philosophy, but for the whole of culture, to be led by science" (Sorell, 1991). Using his theory of transcendental realism, Bhaskar explained that "ontological, epistemological and relational considerations all place limits on the possibility of naturalism" in social sciences. I agree that in contrast to natural sciences, the social sciences cannot be "value-neutral" as they presuppose an axiological commitment to human wellbeing (Bhaskar, 1979) and their subject matter is meaningful action (Weber, 1976). This fact does not exempt social scientists from acting to produce valid theories with explanatory power. For instance, Bhaskar was critical of neoclassical economics, which in his opinion was ill-conceived as a normative theory of efficient action of perfectly rational agents, and was just "generating a set of techniques for achieving given ends, rather than an explanatory theory" (Bhaskar, 1979) for why actual empirical economic phenomena happen. Trying to solve the numerous antinomies (dualities) that social sciences are just referring to but not explaining, such as facts-value, structure-agency, individualism-collectivism, body-soul, Bhaskar developed a critical realist theory of social action which is useful in our attempt to understand why our societies are unsustainable and how we can attempt to free ourselves from the burden of unsustainability. In his transformational model of social activity (TMSA), Bhaskar (1979) started from the premise that societies are real objects irreducible to simpler ones, such as people, given that people and society are different kinds of things. He believed that people do not create society, because society's structures (institutions, conventions, and practices, such as language) exist prior to their birth. However, societies do not exist independently of human activity, given that intentional human agents act – through processes of socialization – to reproduce or transform society's structures (Bhaskar, 1979) (Figure 2.2).

Using the concept of ontic stratification, Bhaskar's TMSA analyzes social events as always taking place in four different spheres (planes): (1) as material transactions with nature; (2) as social interactions between people; (3) as

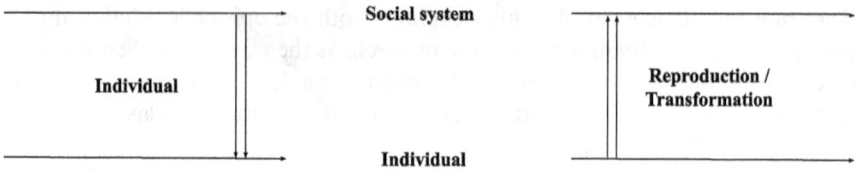

Figure 2.2 Transformational Model of Social Activity
Source: Bhaskar, 1979.

events at the level of social structures *sui generis*; and (4) at the level of the stratified embodied personality. This stratified model of social activity goes directly against the postmodern theory that social life is conceptually constructed, as social life has always had a material dimension, not only a conceptual one. Bhaskar gives the example of war, "which is not just a matter of employing a concept in the correct way, it is the bloody fighting as well" (Bhaskar, 1979). We can give the example of "the insect apocalypse" with, for instance, a 76 per cent decline in insects' biomass over 26 years in Germany from 1989 to 2014, which is not only a dramatic fall in biodiversity due to industrial agriculture, but the real disappearance of moths, butterflies, and bees which are essential for survival of terrestrial and freshwater ecosystems and of the food web (Goulson, 2019). We will use a framework inspired by Bhaskar's TMSA model in the chapter about sustainability in society in order to understand causality in the social world where both human agents and natural and social structures can produce change, by insisting on the role of values in motivating human action.

We have seen so far that in order to study sustainability using critical realism as a theoretical framework, researchers need to have knowledge both about philosophy (ontology) and about epistemology (what we can know about the world and how). They also need to employ a diverse toolkit of research methods to be able to find the best explanation for phenomena or events, not only as observed, or empirically experienced, but in their unseen dimension, as causal powers dominated by the universal laws of nature. This may be a difficult task when empirical data might only exist in the observed event field. Due to the impossibility of the researchers to have direct access to the real domain, research using a critical realism perspective is complex and challenging. Besides expecting the researchers to have disciplinary and interdisciplinary knowledge of the event, it expects them to have a good grasp of the event's context and be able to make inferences as they go deeper in the process of new knowledge creation/theory construction. This inferential type of research attempting to reveal hidden causes will be a continuous interplay of theory development and context analysis, which can be understood as "an iterative, continually interpretive and creative process of analyzing data using different theoretical lenses. Such a process can more clearly reveal generative mechanisms and how they operate" (Ackroyd, 2005). Bhaskar explained that this search for hidden causes should not be seen as metaphysical or less than scientific but a fundamental aspect of causal analysis of social actions, which calls for construction of concepts and theories starting from observed phenomena (Bhaskar, 1978). This explains why critical realism tends to add abductive and retroductive reasoning to deductive and inductive approaches in scientific investigations.

Abductive reasoning is a type of non-deductive logical reasoning, discovered by logician Charles Sanders Peirce at the end of the nineteenth century. He considered it best for inferring conclusions based on observable facts and data through a process of discovering, rather than testing of hypotheses (Peirce, 1960). The researcher using abductive reasoning needs to find the best explanation of an observed phenomenon by developing and reasoning about alternative possible hypotheses. Contrary to deductive reasoning, where an existing theory is used to explain a particular event, abductive reasoning has the potential to create new theoretical knowledge starting from observable facts, as it is "a methodology or a general path toward new knowledge between the empirical and the theoretical realms" (Eriksson, 2015). Also called the method of "inference to the best explanation", an abductive reasoning schemata can take the following form:

ABD1
    Given evidence E and candidate explanations $H_1,..., H_n$ of E, infer the truth of *that H_i* which best explains E.

(Douven, 2021)

The method has been criticized for inferences not being able to lead to real truth, maybe just to "probable" or "likeliest" truth of the best explanation (Lipton, 2004), and for a lack of completeness in the evidence that the data provide (Mukumbang et al., 2020). However, some philosophers believe that abductive reasoning can still be reliable and lead to a true conclusion if "the premises are true" (Douven, 2021). While abductive reasoning is open to the researchers' subjectivity – for instance, in the search for candidate explanations, or in defining the "best explanation" – the inductive inference method is more and more used in social science and medicine realist research, where it is considered a "search strategy which leads us, for a given kind of scenario, in a reasonable time to a most promising explanatory conjecture which is then subject to further test" (Schurz, 2008). For instance, in 2021, David Eriksson and Annika Engström used a research process based on critical realism and abductive reasoning to investigate critical realism's role in operations and supply chain management (OSCM) research. OSCM is a field which is theoretically and philosophically fragmented, and researchers need to use many theories or invent new ones to explain the empirical phenomena of the supply chain. The authors' conclusion was that in spite of the subjectivity involved in the research process, a critical realist approach and abductive reasoning have great explanatory potential, being "inherently cyclical, spiraling toward better and better explanations of generative mechanisms" (Eriksson and Engström, 2021).

Retroduction (from the Latin *retro* meaning "behind" or "beneath") is the inductive reasoning method used for identification of hidden causal forces that lie behind identified patterns or changes in those patterns. It can ask questions such as: "Why do things appear as they do?" (Olsen, 2010), or "What are the necessary conditions for event X to be what it is?" (Danermark et al., 2019), or "What properties do societies possess that might make them possible objects of knowledge for us?" (Bhaskar, 1979). By asking this latter question, Bhaskar assumed that social events take place due to some underlying properties that societies have and he believed that the task of realist research is to uncover and explain those properties, in the attempt to explain potential causes of social change. An inquiry based on critical realism uses both inductive and deductive reasoning, as well as abductive and retroductive theorizing based on empirical data, insights, guesses, or heuristic tools to develop explanatory models in social science. It is thus an appropriate research method to search for hidden causes behind the complex and continually emerging aspects and relationships involved in sustainability. While a research process based on retroductive reasoning is non-formulaic, it can be designed using a stepwise approach. In Thapa and Omland (2018), the following steps are recommended: (1) description of an observed phenomenon and of the empirical events around the phenomenon; (2) identifying the entities and associations (persons, organizations, systems) associated with the phenomenon, and collecting data about these entities; (3) interpreting the data, searching for different theoretical perspectives and different explanations – abduction; and (4) hypothesizing about the mechanisms and conditions that might have produced the phenomenon – retroduction. A last step would be to compare the explanatory power of the developed hypothesis to other competing hypotheses. These steps show that abduction and retroduction are central to critical realist-informed research, as they are the procedures able to move research from a description of a social phenomenon to an explanatory theory of that phenomenon, and create questions for traditional scientific research to pursue. In spite of their highly explanatory potential in realist social science research, abductive and retroductive reasoning are seldom explicitly applied and described in such studies which are dominated by deductive and inductive logical approaches in the process of theory formulation (Mukumbang et al., 2021). This is unfortunate, as it could be that critical realism and the inference methods at its core are the best coherent framework we have to investigate complex topics such as sustainability, or what people think and know about these issues. I argue that critical realism allows us to use a framework that can help prevent the complex interdisciplinary areas of study necessary for sustainability from becoming "a chaotic mix of social interpretation and physical facts" (Bhaskar, 2010). As such, I intend to use abductive reasoning in the chapter "Who is the Sustainer?" as my attempt to find the best possible explanation about the incredible force behind sustainability.

This chapter has clarified why understanding "sustainability" not only as a concept but as an objective feature of the Universe requires a richer ontology and epistemology, and the subsequent choice of research methods rooted in Popper's realism, explaining the "propensities" we observe in the world, and in Bhaskar's critical realism which sees the world as stratified and including a "real" dimension which can embed transcendental aspects as causes of phenomena.

References

Ackroyd, S. (2005) "Methodology for management and organisation studies: some implications of critical realism." In *Critical Realism Applications in Organisation and Management Studies*, edited by S. Fleetwood and S. Ackroyd. London: Routledge, pp. 142–165.

Archer, M. (2007) "The ontological status of subjectivity: The missing link between structure and agency." In *Contributions to Social Ontology*, edited by C. Lawson, J. Latsis, and N. Martins. London: Routledge, pp. 17–31.

Berger, P.L. and Luckmann, T. (1966) *The Social Construction of Reality: A Treatise in the Sociology of Knowledge*. New York: Doubleday.

Bhaskar, R. (1975) *A Realist Theory of Science*. London: Routledge.

Bhaskar, R. (1978) "On the Possibility of Social Scientific Knowledge and the Limits of Naturalism." *Journal for the Theory of Social Behaviour*, 8(1): 1–28.

Bhaskar, R. (1979) *The Possibility of Naturalism: A Philosophical Critique of the Contemporary Human Sciences*. Atlantic Highlands, NJ: Humanities Press.

Bhaskar, R. (1989) *Reclaiming Reality*. London: Verso.

Bhaskar, R. (2009) *Scientific Realism and Human Emancipation*. Abingdon: Routledge.

Bhaskar, R. (2010) "Contexts of Interdisciplinarity." In *Interdisciplinarity and Climate Change*, edited by R. Bhaskar, C. Frank, K.G. Hoyer, P. Naess, and J. Parker. London: Routledge, pp. 15–38.

Bohm, D. (1980) *Wholeness and the Implicate Order*. London and New York: Routledge.

Carroll, S. (2017) *The Big Picture: on the Origins of Life, Meaning, and the Universe Itself*. New York: Dutton.

Creswell, J.W. and Miller, D.L. (2000) "Determining Validity in Qualitative Inquiry." *Theory into Practice*, 39: 124–130.

Danermark, B., Ekström, M., and Karlsson, J.C. (2019) *Explaining Society: Critical Realism in the Social Sciences*, 2nd ed. London: Routledge.

Douven, I. (2021) "Abduction." In *The Stanford Encyclopedia of Philosophy*, edited by E.N. Zalta. https://plato.stanford.edu/archives/sum2021/entries/abduction.

Eriksson, D. (2015) "Lessons on knowledge creation in supply chain management." *European Business Review*, 27(4): 346–368.

Eriksson, D. and Engström, A. (2021) "Using critical realism and abduction to navigate theory and data in operations and supply chain management research." *Supply Chain Management*, 26(2): 224–239. https://doi.org/10.1108/SCM-03-2020-0091.

Farley, H.M. and Smith, Z.A. (2020) *Sustainability: if it's everything, is it nothing?* 2nd ed. New York: Routledge.

Gale, F.P. (2018) *The Political Economy of Sustainability*. Cheltenham, UK and Northampton, MA: Edward Elgar.

Goulson, D. (2019) "The insect apocalypse, and why it matters." *Current Biology*, 29: R967–971.

Gorski, P.S. (2013) "What is Critical Realism? And Why Should You Care?" *Contemporary Sociology*, 42(5), 658–670. https://doi.org/10.1177/0094306113499533.

Lipton, P. (2004) *Inference to the Best Explanation*, 2nd ed. London: Routledge.

Lorenz, E.N. (1963) "Deterministic nonperiodic flow." *Journal of the Atmospheric Sciences*, 20: 130–141.

Mukumbang, F.C., Marchal, B., Van Belle, S., and van Wyk, B. (2020) "Using the realist interview approach to maintain theoretical awareness in realist studies." *Qualitative Research*, 20(4), 485–515. https://doi.org/10.1177/1468794119881985.

Mukumbang, F.C., Kabongo, E.M., and Eastwood, J.G. (2021) "Examining the Application of Retroductive Theorizing in Realist-Informed Studies." *International Journal of Qualitative Methods*, 20: 1–14.

Nuijten, E. (2011) "Combining research styles of the natural and social sciences in agricultural research." *NJAS – Wageningen Journal of Life Sciences*, 57(3–4):197–205.

O'Hare, A. (1985) "Critical Notices." *Mind*, XCIV(375): 453–471. https://doi.org/10.1093/mind/XCIV.375.443.

Olsen, W. (2010) "Editor's Introduction: Realist Methodology – A Review." In *Realist Methodology: Benchmarks in Social Research Methods Series*, edited by W. Olsen. Los Angeles: Sage.

Peirce, C. (1960) *Collected papers of Charles Sanders Peirce*. Harvard, MA: Belknap Press of Harvard University Press.

Popper, K. (1945) *The Open Society and Its Enemies* (2 volumes). London: Routledge.

Popper, K. (1959) The Logic of Scientific Discovery. London: Hutchinson. [English translation of *Logik der Forschung*, Vienna: Springer (1935).]

Popper, K. (1972) *Objective Knowledge: An Evolutionary Approach*. Oxford: The Clarendon Press.

Popper, K (1990) *A World of Propensities: Two New Views of Causality*. Bristol: Thoemmes Antiquarian Books.

Quraishi, M., Irfan, L., Schneuwly Purdie, M., and Wilkinson, M.L.N. (2022) "Doing 'judgmental rationality' in empirical research: the importance of depth-reflexivity when researching in prison." *Journal of Critical Rationalism*, 21(1): 25–45. https://doi.org/10.1080/14767430.2021.1992735.

Ratner, B.D. (2004) "'Sustainability' as a Dialogue of Values: Challenges to the Sociology of Development." *Sociological Inquiry*, 74(1): 50–69.

Sabau, G.L. (2001) *Societatea cunoasterii. O perspectiva romaneasca [Knowledge Society. A Romanian Perspective]*. Bucuresti: Editura Economica.

Schurz, G. (2008) "Patterns of Abduction." *Synthese*, 164: 201–234.

Smith, S.P. and Johnston, R.B. (2014) "How Critical Realism Clarifies Validity Issues in Information Systems Theory-Testing Research." *Scandinavian Journal of Information Systems*, 26(1): Article 1. http://aisel.aisnet.org/sjis/vol26/iss1/1.

Sorell, T. (1991) *Scientism Philosophy and the Infatuation with Science*. London and New York: Routledge.

Thapa, D. and Omland, H.O. (2018) "Four steps to identify mechanisms of ICT4D: A critical realism-based methodology." *The Electronic Journal of Information Systems in Developing Countries*, 84(6): e12054. https://doi.org/10.1002/isd2.12054.

Weber, M. (1976) *The Protestant Ethic and the Spirit of Capitalism*. London: George Allen & Unwin.

Westhorp, G. (2013) "Developing complexity-consistent theory in a realist investigation." *Evaluation*, 19(4): 364–382.

3 Sustainability as objective physical reality

We live in a spectacular Universe which is both a mysterious place and a complex, intricate system functioning as "an interconnected set of elements that is coherently organized around some purpose" (de Vries, 2012) or "function" (Meadows, 2008). The elements are planets, stars, moons, comets, galaxies, and clusters of galaxies, orbiting around in an orderly fashion and held together by the force of gravity, but also huge amounts of dark energy and matter which do not interact electromagnetically with light. The Universe exists objectively (apart from our conscious observation of it) as "astronomy educates us that the motions of planets and stars have nothing to do with human actions" (Loeb, 2020). However, the Universe can be known both by observation and by scientific probing, both at the micro and the macro levels, and over the last century, scientists have managed to unveil many of the Universe's mysteries. We know, experientially, as alive inhabitants of planet Earth, that our Universe exists and functions as a vast, ordered, complex, dynamic, and lasting life-support mechanism available to all living things on our planet. This feature of the Universe, which is called "sustainability" in this book, implies that life in all forms is precious, it is worth sustaining, and that humans, as the only living things on Earth endowed with reason and "the moral law within" (Kant, 2004), are expected to participate responsibly in the life-web at the core of sustainability, obeying the pre-set rules of functioning of the physical Universe. How did sustainability come to be on our life-friendly planet? Is it the result of just blind evolutionary processes (Darwin, 1860) or is there an unfolding purpose in our life-friendly Universe, which seems to behave as if it "must have known we were coming" (Dyson, 2001)? The goal of this chapter is to show, based on scientific evidence from physics, astronomy, chemistry, cosmology, and biology, that sustainability exists objectively in nature and is not the result of blind forces, as it is embodied in structures, functions, and processes specifically designed to protect and sustain life. The research approach is based in critical realism with its two important principles: that there is a real world "out there" which exists independently of humans, and that this world can be investigated by science through a spiralling process of discovery which can take us closer to understanding the essence of sustainability in nature. On this view, "science is not an epiphenomenon of nature, nor is nature a product of man" (Bhaskar, 1975).

DOI: 10.4324/9781003307587-4

The fact that our Universe is "comprehensible" is "the eternal mystery of the world", wrote Einstein in 1936 (Einstein, 1936/1993) hinting to the fact that some higher spirit must be behind the order we perceive with our rational minds in our complex Universe. The assumption that the world is intelligible led in the natural sciences since the sixteenth century to "extraordinary discoveries, confirmed by prediction and experiment, of a hidden natural order that cannot be observed by human perception alone" (Nagel, 2012). Famous astronomers and physicists such as Nicolas Copernicus (1473–1543), Galileo Galilei (1564–1642), Isaac Newton (1643–1727), and many others have dedicated their lives to understanding the hidden natural order of the Universe and to grasping the "language and symbols in which it is written" (Galilei, 1618/1960), thus paving the way to modern science. It was Sir Isaac Newton who first described how the macroscopic objects in the observable Universe behave, in his classical mechanics developed in the book *Philosophiae Naturalis Principia Mathematica* (Newton, 1687/1962). Other eminent physicists in the late nineteenth century and early twentieth century were trying to understand the behaviour of matter and energy at the tiniest particle level. They discovered that light and subatomic particles (electrons, protons) behave simultaneously both like particles with mass and like waves without mass. In 1900, the German physicist Max Planck discovered that matter emits energy in small discrete amounts called "quantas", and that is how the field of quantum mechanics was born. Planck's constant relates the energy in one quantum (photon) of electromagnetic radiation to the frequency of that radiation. In 1905, Albert Einstein discovered the equivalency of energy and matter, showing that the mass of a system measures its energy content according to his famous formula $E=mc^2$. This discovery allowed Einstein to offer a new explanation for gravity, in his general relativity theory which states that the force of gravity arises from the curvature of space-time under the influence of massive bodies rather than from an attraction force between masses, as Newton's classical physics had previously described universal gravity. In 1926, the Austrian physicist Erwin Schrödinger derived the fundamental equation of quantum mechanics, an equation describing the dual nature of light and subatomic particles. His equation allows physicists to calculate the wave function that governs the probable motion of subatomic particles, and how their location and movement are altered by external influences (Britannica, n.d.). Further research by Schrödinger has shown that entanglement of particles and systems – either microscopic or macroscopic – is "the characteristic trait of quantum mechanics" (Schrödinger, 1935), and it is not due to some "local hidden variables" as Einstein and his team, for instance, maintained in a famous 1935 article (Einstein et al., 1935). Schrödinger's insight was confirmed in 2022 by the findings of three Nobel prize physicists who demonstrated experimentally that two entangled particles, travelling separated, are able to re-entangle, even at cosmic distances, by adjusting their spin as if they benefitted from information encoded in their quantum states (Nobel Prize in Physics, 2022). In 1927, the American astronomer Edwin Hubble discovered that the Universe is expanding. The rate of expansion, known as the Hubble constant, is an important figure that "yields clues to the origin, age, evolution, and future fate of our universe" (New Hubble, 2019). There is a consensus among scientists that the Universe had a beginning in a "big bang" explosion, a theory confirmed in 1965 by the discovery of the low-energy cosmic background radiation left over from the initial "big bang" explosion (Singh, 2005). The discovery that the Universe had a beginning implied that it also has an end, and it allowed astronomers to calculate the approximate age of the Universe. It is believed that the Universe has been around for about 13.8 billion years (Planck, 2015), and has functioned quite predictably due to a number of constants, fundamental initial boundary conditions built into its structure from the beginning. Up to now, scientists have identified at least 31 fundamental constants (Tegmark et al., 2006) which keep the Universe "predictable and stable" (Giberson, 2012). These constants do not include the values of the four fundamental forces of nature: the electromagnetic force, the force of gravity, the strong nuclear force (an attractive force holding protons and neutrons together), and the weak nuclear force (the force causing nuclear radiation through radioactive decay of atoms). The values of the fundamental constants are non-negotiable; they must be accepted as brute facts of the Universe. For instance, the value of acceleration due to gravity is $9.81 \ m/(s^2)$, where m is the object's mass and s is the time in seconds. This means that a free-falling object with mass m will experience this rate of increase in velocity per second s under the influence of gravity. There are some constants in the Universe, such as the cosmological constant (a measure of the energy density of the vacuum), whose behaviour cannot be currently explained by science, as seen through the relatively recent discovery that the expansion of the Universe is accelerating (Perlmutter et al., 1999; Riess et al., 1998). Astronomical observations indicate that the cosmological constant is many orders of magnitude smaller than it was estimated in modern theories of elementary particles (Weinberg, 1989), as if, indeed, "a super-intellect has monkeyed with physics, as well as with chemistry and biology" (Hoyle, 1982) by preventing the vacuum energy from gravitating to the full extent. Einstein did not like such "surprises" in the Universe, which he wanted to be deterministic and able to be measured by clear laws of physics representing a time-space reality, "free from spooky action at a distance" (The Born-Einstein Letters, 1971). However, recent research in quantum physics shows that there is "spooky action at a distance" in our probabilistic Universe where, due to quantum entanglement, "the possibility to 'teleport' an arbitrary quantum state from one position to another remains, so long as the original copy is destroyed" (Nobel Prize in Physics, 2022).

Is there purpose in nature?

Due to the probabilistic character embedded in the quantic structure of our Universe, it is difficult to identify purpose in nature. It is easier to assume that nature is in principle completely knowable, based on nature's regularity, unity, and wholeness that implies objective laws and no supernatural events or entities (Britannica, n.d.). This is what the theory of naturalism holds, that all reality is natural and knowable by science, when science studies it in well-ordered disciplinary entities. The theory of naturalism was built gradually in Western thought – starting at the beginning of the seventeenth century – as scientific advances in astronomy, medicine, and cosmology led to the unravelling of the medieval narrative accepted for centuries about a teleological unitary cosmos created by a rational super-intellect which sustains it, giving a privileged place to humans. This view is also promoted by the monotheistic religions – Christianity, Islam, and Judaism – with their belief that the world is the result of a supernatural cause which must be a rational one, as it is "not blind and fortuitous, but very well skilled in Mechanicks and Geometry", as Newton famously noted (Cohen 1978). Teleological notions are also defended in the Mayan, Zuni (New Mexico Indian), the "Thompson" Indian of the North Pacific coast, Iroquois, Sumerian, Bantu, ancient Egyptian, Islamic-Persian, and Chinese traditions (Barrow and Tipler, 1986). For instance, in the Iroquois tradition, the Earth was created for the benefit of human beings by the Sky god Sapling who placed humans on the Earth "for the purpose that they shall continue my work of creation by beautifying the Earth, by cultivating it and making it more pleasing for the habitation of man" (Barrow and Tipler, 1986). In the same way, some Sumerian creation legends show that gods created humans to serve them by worshipping and offering sacrifices, but also to "imitate them in creating and preserving the cosmic order" (Eliade, 1978). Views about wholeness of the world still survive in the East (particularly in India) where "philosophy and religion emphasize wholeness and imply the futility of analysis of the world into parts", mainly based on the insight that it is impossible to measure "the immeasurable (i.e. that which cannot be named, described, or understood through any form of reason)" (Bohm, 1980).

Naturalism became a dominant theory in philosophy and science in the age of the Enlightenment during the seventeenth and eighteenth centuries in Europe, an age dubbed as "the age of reason", or of human understanding of the Universe independently from any teleological presuppositions of traditional medieval or theistic thinking. Philosophers such as Hume, Comte, Kant, Laplace, and others "disenchanted" nature by seeing the Universe as functioning autonomously, moved by "a chain of natural causes [which] would account for the construction and preservation of the wonderful system" (Kaiser, 1991). These philosophers contributed to the spread of a secular cosmology, which excluded all teleological and design inferences from nature. The Scottish philosopher David Hume was a sceptical empiricist. He rejected the rationality and the ordered structure of the Universe, and believed that the laws of nature were established through human senses' "firm and unalterable experience" (Hume, 1777/1902). However, Hume admitted that matter possessed some mysterious ability to self-organize: "such is the nature of material objects and that they are originally possessed by a faculty of order and proportion" (Hume, 1777/1902). Likewise, Auguste Comte, the father of positivism, maintained that real, "positive" knowledge of natural phenomena can only be attained when scientists replaced superstitions and philosophical abstractions about final causes with natural laws and strictly material mechanisms (Comte, 1858). The German philosopher Immanuel Kant, in his *Universal Natural History*, saw purpose and design in nature. He believed that the goal of nature is biodiversity, the goal of planets is to sustain biospheres and intelligent life, and the goal of terrestrial existence is to increase biota. Kant concluded that "the telos of nature is life", at least until it reaches maximum density and begins to fracture in cosmic collapse (Kant, 1755/2012). Also, in *The Critique of Judgement*, Kant acknowledged that mechanistic natural laws cannot explain the genesis of "organised beings and their inner possibility", not even for "a blade of grass […] that no design has ordered" (Kant, 1790/1952). In later works, however, Kant undermined this teleology by expressing scepticism concerning the Kalam cosmological argument, according to which the physical Universe had a beginning being created by an uncaused First Cause (Craig, 2011). Kant argued that we cannot identify final or first causes in the real world by pure reason alone (Kant, 1787/1968). Of course, Kant was right, as by pure reason we cannot perceive the immeasurable aspects of an unseen final cause. Final cause refers to Aristotle's philosophy of causation, which included four types of causes explaining how any being came to be. These causes are material, efficient, formal, and ultimate. If we think of a tree, the material cause is the wood, the soil, water, and solar energy that contribute to the material body of the tree. The efficient cause is an action, external to the tree, such as the planting of a seed, that brought the tree into being. The formal cause is a "formative cause", or the "whole inner movement of sap, cell growth, articulation of branches, leaves, etc., which is characteristic of that kind of tree and different from that taking place in other kinds of trees" (Bohm, 1980). The final cause is the end or purpose that the tree will serve, such as to provide shade in a garden or to be part of a larger cosmical whole of living beings, such as a forest. The final cause is most of the times *implicit* in the whole process, and humans can be made aware about what it is only when they see their activity of thinking about reality as "a form of insight, a way of looking, rather than as a 'true copy of reality as it is'" (Bohm, 1980). Bohm believed that seeing the reality as fragmented and not as whole has led in our modern civilization to the numerous crises that humanity experiences. I concur with Bohm's assessment and believe that the confusion about the numerous dualities we introduced in our thinking has prevented humans from seeing sustainability in our Universe for what it is.

The doctrine of final causes, and all traditional teleology, was rejected in 1859 by Charles Darwin in his book *On the Origin of Species by Means of Natural Selection, or the Preservation of Favoured Races in the Struggle for Life*. In this book, Darwin proposed a new mechanism for explaining the adaptation of biological organisms to their environment. The mechanism was based on natural selection through random variations and inheritance. According to this theory, all biological species have a common ancestor, and natural selection leads to individuals better adapted in the struggle for existence, or to "survival of the fittest" (Spencer, 1864). Darwin's theory was based on a vast amount of adaptation evidence collected by Darwin from nature between 1831 and 1836, but it did not explain the origin of the first living cell, nor the mechanism of inheritance. In the *Origin of Species*, Darwin did not apply his theory of evolution to human species, but he did so in the book *The Descent of Man*, published in 1871. With this book, according to which humans have evolved from lower animals through the mechanism of natural selection, "teleology, as commonly understood, had received its death-blow at Mr. Darwin's hands" (Huxley, 1871). Darwin's theory of evolution was controversial from the beginning. For instance, the British physicist Lord Kelvin questioned the soundness of the theory by arguing that the geophysical evidence pointed towards a terrestrial age too brief for natural selection to evolve the observed spectrum of living creatures (Barrow and Tipler, 1986). However, Darwin's theory of natural selection became very influential and was established as the undisputed scientific explanation for evolution of the cosmos and of life on Earth in the last 160 years. The influence of Darwin's theory of evolution was extended beyond the sphere of biology and became the unquestioned scientific dogma for social scientists who introduced naturalistic thinking in social, economic, and racial theories.

The discovery in 1944 by the biologist Oswald Avery and his team that desoxyribonucleic acid (DNA) is the physical substance of genes and heredity in living cells paved the way for discovery in 1953 by Francis Crick and James Watson of the double-helix structure of DNA which could store and transmit codified information. This discovery made molecular biologists and geneticists question the ability of Darwin's theory of evolution to explain the numerous design-like structures and functions in the living cell, especially the source of the codified information in the DNA molecule (Denton, 1985). In 1991, in his book *Life Itself: A Comprehensive Inquiry into the Nature, Origin, and Fabrication of Life,* mathematician Robert Rosen criticized biologists for failing to explain the highly complex systems of living organisms by believing that "everything important about life is not necessary but contingent" (Rosen, 1991). In addition, applied biologists who discovered the role of fine-tuning in optimizing gene networks, noted the difference between Darwin's random evolution and directed evolution (Egbert and Klavins, 2012). In 2004, the famous biophysicist and microbiologist Carl Woese called biologists "to imagine a biology released from the intellectual shackles of mechanism, reductionism, and determinism", one "with a new and genuinely holistic, 'eyes-up,' view of the living world, one whose primary focus is on evolution, emergence, and biology's innate complexity" (Woese, 2004). He warned biologists that "[s]ociety cannot tolerate a biology whose metaphysical base is outmoded and misleading: the society desperately needs to live in harmony with the rest of the living world, not with a biology that is a distorted and incomplete reflection of that world" (Woese, 2004). Woese was referring to the epistemological error of considering the living organisms as just molecular machines, as "while a machine is a mere collection of parts, some sort of 'sense of the whole' inheres in the organism" (Woese, 2004).

These evolutions led, among more and more biologists, to the need for a rethinking of the Darwinian theory of evolution. In 2014, a group of molecular biologists, convinced that the drivers of evolution are not the genes only, proposed replacing Darwin's theory of evolution with a broader framework, termed the "extended evolutionary synthesis" (EES) (Laland, 2014), one able to give explanations about new phenomena in molecular biology such as plasticity, evolutionary development, epigenetics, and cultural evolution. A June 2022 article entitled "Do we need a new theory of evolution?" contains a new call of a group of biologists for "a new way of thinking about evolution – one that starts not by seeking the simplest explanation, or the universal one, but what combination of approaches offers the best explanation to biology's major questions" (Buranyi, 2022).

The Darwinian revolution is of interest for our discussion about sustainability, as its idea of unguided evolution perpetuates the false and dangerous myth that "ecosystems will in time adapt to the changes we inflict upon them" (Johnston et al., 2007). This idea dismisses our moral responsibility for overusing ecosystems' functions and services. In addition, if we believe that there is no purpose in the Universe, the very idea of sustainability which implies that there is "an impulse to life implanted in all of us" (MacIntyre, 1981) is undermined. And if humans are the un-purposed by-products of blind biological evolution, they cannot be expected to make sustainable choices or be held accountable for their unsustainable ways. For "if mechanism reigns in nature, it reigns everywhere, and in ethics as well as in physics" (Janet, 1878). However, we cannot ignore the fact that humans recognize the moral law within, which speaks about the potential existing in all humans to act morally, as if that has been the goal or *telos* for which they exist. "Morality is, therefore, at once the accomplishment and the ultimate proof of the law of finality" (Janet, 1878).

Nature is fine-tuned for life

Even if the *telos* in nature is not easily deciphered by using the deductive reasoning methods dominant in natural sciences, still scientists can develop hypotheses concerning the causes of various strange concurrences of events or natural phenomena which, when tested, can reveal a purpose. One such hypothesis is that planet Earth, the part of the Universe where biological life exists, is fine-tuned to support life, where fine-tuning refers to the precise balancing of the fundamental laws and parameters of physics, chemistry, and biology and of the initial matter and energy arrangements in the Universe to make life possible and to sustain it. Can we infer that our carbon- and oxygen-rich Universe was purposefully built to sustain the life of oxygen-breathing, carbon-based beings like us? Some scientists believe so when they observe:

[I]n the set of fundamental parameters (constants and initial conditions) of nature, such as the cosmological constant and the strength of electromagnetism, an extraordinarily small subset would have resulted in a universe able to support the complexity required by life.

(Barnes, 2020)

The first references to the fitness of nature for life were made in the Bridgewater Treatises of the 1830s, where authors discussed the special properties of water and of carbon atoms to support life on Earth (Denton, 2022). At the beginning of the twentieth century, Lawrence Henderson, in his classic book *The Fitness of the Environment*, explained, using knowledge of organic chemistry, how the properties of matter and energy, as well as the course of cosmic evolution are specifically fit to support carbon-based life on our planet (Henderson, 1913). He pointedly referred to water and the compounds of carbon, hydrogen, and oxygen, such as carbon dioxide, and their strange fitness for biological life. He concluded that biologists should explore more the natural Universe for finding explanations about biological life, as he felt that there is unity between the make-up of the cosmos and biological life on Earth. In his words, "For the whole evolutionary process, both cosmic and organic is one, and the biologist may rightly regard the universe in its very essence as biocentric" (Henderson, 1913). Henderson's conclusion is interesting for our sustainability research, as it describes a Universe which is a unified whole, with the life-supporting attributes "built in" the natural order from the beginning. The picture it paints is so different from Darwin's idea that living organisms needed to adapt by chance to their environment in order to evolve higher-level life forms. In spite of its popularity, the theory of evolution, and its two main concepts of fitness and natural selection, provide a very poor explanation of biological life's continuity, as the anecdotal examples of evolution, such as the peppered moth experiments, "provide no real evidence that Darwinian mechanisms are at work and no evidence, surely, that Darwinian theories *explain* the process by which a new species arises from one that is old" (Berlinski, 2023, quoting Wells, 2017).

Henderson's interest in carbon was due to the fact that all life on Earth is carbon-based, and to the observed property of carbon to bond easily with other elements, by being "in fact the *only* molecule around which we can build the incredible complex molecules – sugars, acids, proteins – that make life possible" (Giberson, 2012). Carbon also combines with oxygen to form carbon dioxide, a compound essential for internal respiration in living bodies, for photosynthesis, and for storing food in plants. How do we explain that carbon is prevalent in life forms in our Universe, when carbon is produced inside stars from the fusion of highly unstable atoms of beryllium and helium? The fusion would only occur if a version of higher energy state (resonance) carbon-12 nucleus existed, hypothesized astrophysicist Fred Hoyle. Hoyle calculated that the necessary carbon resonance level for the fusion to take place must be 7.65 MeV (million electron volts) above its ground-state

(Burbidge, 1957). The accuracy of Hoyle's calculation was confirmed through empirical testing in 1935 by astrophysicist William Fowler, who brought another confirmation that the Universe is fine-tuned to support life on Earth. In 2000, in his book *Just Six Numbers: The Deep Forces that Shape the Universe*, British Royal Astronomer Sir Martin Rees explained how just six basic physical constants or their ratios, imprinted in the "big bang", determine the essential features of the physical cosmos, making life on Earth possible. He also showed that cosmic evolution is highly sensitive to the numerical values of these constants and that if any one of them were not tuned, there could be no stars and no life, "as we know it" (Rees, 2000). In 2016, Lewis and Barnes, in their book *A Fortunate Universe: Life in a Finely Tuned Cosmos*, noted that the "universe seems inexplicably well-tuned to facilitate the evolution of complex molecular structures and sentient creatures" (Lewis and Barnes, 2016). They clearly talk about sustainability in nature when they refer to various features of the Earth which are able to sustain the existence of life at the level of "a single, 'simple' cell" which is "a miracle of complexity" (Lewis and Barnes, 2016). More recently, the Brazilian chemist Marcos Eberlin, in his book *Foresight: How the Chemistry of Life Reveals Planning and Purpose*, discussed numerous examples from his mass spectrometry laboratory research which indicate that the chemistry of life provides ingenious solutions to protect and accommodate life in biological systems (Eberlin, 2019). Eberlin's findings have been confirmed in a recent article describing molecular fine-tuning in biological systems at different levels, in functional proteins, in complex biochemical machines existing in living cells, and in cellular networks (Thorvaldsen and Hössjer, 2020).

A critical realist view of sustainability in nature

We have seen so far that there have been impressive advances in natural sciences describing our Universe and the way it functions. Numerous discoveries in physics, astronomy, chemistry, biology, and scientific cosmology show that the Universe is a unified, orderly, structured system which functions both according to predictable natural laws, which are manifestations of specific strengths of various fundamental forces of nature, and according to the unpredictable ways of quantum mechanics. The "Big Bang" cosmological model, based on Einstein's general theory of relativity, predicts that the Universe is also dynamic (expanding), and is big and old enough to allow life, including intelligent life, to exist, develop, and continue. The problem of whether this life-sustaining configuration of the Universe was designed from the beginning or not is a controversial one among scientists, cosmologists, and theologians. The mainstream science position is that life sustainability is the result of natural evolution. But more scientists' voices, especially from mathematics, microbiology, medicine, and genetics, are challenging this worldview, considering that "randomness alone is incapable of creating the living world as we know it. [...] You simply cannot create something so intricate as the interdependent, interconnected biosphere with its vastly complex creatures by spinning a roulette wheel" (Iacoboni, 2022). The limits of natural selection have been acknowledged by evolutionary biologist Andreas Wagner in his book *The Arrival of the Fittest: How Nature Innovates*:

> Natural selection can preserve innovations, but it cannot create them. [...] Nature's many innovations – some uncannily perfect – call for natural principles that accelerate life's ability to innovate. [...] For the last fifteen years, I have been privileged to help uncover these principles. [...] What we have found tells us that *the principles of innovation are concealed*, even beyond the molecular architecture of DNA, *in a hidden architecture of life* with an otherworld beauty.
>
> (Wagner, 2014; emphasis added)

I will try to weigh in on this issue by looking at our Universe and what we know about it using the realist theory of science (Bhaskar, 1975). This theory sees reality as stratified in three distinct domains: the real, the actual, and the empirical. By viewing reality in this way, it is possible to explain sustainability not only through empirical experiences in nature, but also through actual events and phenomena, as well as unseen mechanisms or forces that produce effects in nature.

Empirically, we know that we live in a life-friendly Universe where we have air to breathe, potable water to drink, waterways for travel, starry skies at night, and rainforests teeming with biodiversity. We count on summer to follow spring, on the sun to rise every morning, and on soil to give us good crops if we get the right amount of rain and sunshine when needed. We take for granted the fact that we inhabit an infinite Universe which is comparatively isotopic and homogeneous, and are surprised when it snows in June in places where it shouldn't, or when a hole forms in the ozone layer over Antarctica due to too much anthropocentrically produced pollution. Life unfolds in a predictable pattern: all living beings are born, grow, mature, and die without exception, leaving behind living legacies that allow life to continue on to next generations. Sometimes, this pattern is interrupted by unexpected events, such as tornadoes, earthquakes, wildfires, global pandemics, or miraculous healings, which all remind us that we do not run the nature-aggregate which is probabilistic not only at the quantum-mechanics level. Scientific observation can help us probe more deeply into the Universe, aided by powerful telescopes and microscopes, and to discover the *actual reality* of many science-confirmed life-sustaining features of our planet. And yet, even when using powerful tools of scientific investigation, we can only observe a finite portion of our Universe, "on and inside our past light-cone, which is defined by that set of signals able to reach us over the age of the universe" (Barrow and Tipler, 1986). According to best available science, based in biometric dating of terrestrial rock formations, the Earth has existed for approximately 4.55 billion years (Patterson, 1956) and has provided life-sustenance in the earliest habitable environments, the submarine-hydrothermal vents, which have harboured low forms of life (bacterial cells) for more than 3.77 billion years (Dodd et al., 2017). It is assumed that *Homo sapiens* has been around for about 286,000 years (Richter et al., 2017), and that approximately 100 billion humans lived on the Earth so far (Sinclair, 2019). The Earth is bathed in a continuous flow of solar energy, which is essential for plants and animals to live and reproduce. Every year, the Sun sends the Earth 5.6310×10^{24} J of energy, which is over 10,000 times more energy than humans consume in a year (Nielsen at al., 2020). The Sun's electromagnetic radiation provides the Earth with heat and light, regulates climate and the hydrological cycle and, through photo-synthesis, makes plants grow and store energy which powers either biological bodies or is accumulated in geological time in fossil fuels. Being situated at the right distance from the Sun, the Earth is the only known planet in the Universe which has liquid water, which is critical for life and an ideal solvent for most biochemical reactions. Water has numerous miraculous life-sustaining proper-ties such as being less dense when it is frozen (ice floats on water). It reaches maximum density at 4 degrees Celsius, which allows lakes and seas to freeze at the surface while protecting marine life in the deep (Denton, 2017). The Earth's atmosphere has the right composition (nitrogen, oxygen, argon, and trace amounts of carbon dioxide and water vapour), temperature range, and pressure for supporting life and protecting it from dangerous radiation from space. The ozone layer, situated in the lower part of the stratosphere, contains exactly the concentration required to block harmful ultraviolet radiation from the Sun and to allow a useful amount of beneficial ultraviolet radiation to pass through (Eberlin, 2019). Nature on Earth is bountiful, resilient, and orderly (as seen through the food web or the periodic table of elements, for instance). There is competition but also cooperation and mutual dependence among living species within ecosystems, as well as network synergies (Fiscus and Fath, 2019), which suggests that "this is more than the simple outworking of some survival equa-tions from game theory. Our deeply rooted affinity for community is a part of the way things are supposed to be" (Giberson, 2012). Japanese researchers have discovered that tree leaves fallen into streams and rivers leach acids into the ocean that stimulate the growth of plankton, the first building block of the food web, thus promoting growth of fish and oysters in the area (Robbins, 2012). A well-researched mutually beneficial relationship exists between trees and mush-rooms growing at the foot of the trees. The mushrooms absorb sugars and other carbohydrates from photosynthesis occurring in tree's leaves while the trees benefit from essential minerals and increased water uptake. "You find twice the amount of life-giving nitrogen and phosphorus in plants that coop-erate with fungal partners than in plants that tap the soil with their roots alone" (Wohlleben, 2015). But nature is also frugal. There are efficient produ-cers and decomposers in nature, and numerous creatures are in the business of recycling; nature's organic waste does not pile up, as scavengers and micro-organisms dispose of organic waste, get fed, and improve the quality of soil in the process (Gunderson and Holling, 2009).

On the other hand, the Earth system has some clear boundaries which are significant for sustainability. Century-old sequoia trees do not grow forever in height; they grow to be about 50 metres tall with a diameter of 1.8 metres at chest height (Wohlleben, 2021). Likewise, puppies and children do not grow endlessly; they have growth mechanisms built into their DNA and stop growing at the predetermined time. Hunger and thirst can be satiated, as the law of diminishing marginal utility teaches. There is a very significant macro boundary which is determined by the fact that the Earth is a "mostly non-isolated system, exchanging with the environment energy and information, but only incidentally mass" (Nielsen et al., 2020), such as when a meteorite lands on our planet. While we count on a constant flow of solar energy, which is assumed to be available for at least 5 billion more years (Giberson, 2012), there are absolute limits to the amount of matter and energy available on Earth, with limits indicated by the laws of thermodynamics (Clausius, 1865). According to the first law of thermodynamics, we cannot create more matter and energy (the law of conservation). The second law states that the entropy, "a direct measure of molecular disorder" (Schrödinger, 1944), tends to increase in isolated systems, such as the Universe. Such systems tend to irreversibly move from a more orderly state to a more random state, ending in thermodynamic equilibrium (death). This is important for sustainability, as the solar energy flow and biogeochemical cycling set an upper limit on the quantity and number of organisms and on the number of trophic levels that can exist in an ecosystem (Odum, 1977). These limits should guide the level of economic activity, given that "all natural resources are irreversibly degraded when put to use in economic activity" (Georgescu-Roegen, 1971). Ecological economics has described these objective limits in the vision of the circular flow of economic activity embedded in the Earth's ecosystem (Figure 3.1), so different from the circular flow diagram of goods and money described in neoclassical economics, where the economy exists on its own (Figure 3.2). Natural scientists have identified some of these limits (Vitousek et al., 1997; Wackernagel and Rees, 1995; Millennium Ecosystem Assessment, 2005; Rockström et al., 2009; Lenton et al., 2019) and have warned that we live in a human-dominated planet which we are changing more rapidly than we understand it. Achieving sustainability requires not only a thorough knowledge of the planet's diverse ecosystems in their "natural" states, but also rigorously treating their structure and functions "as a fixed reference point to frame our development activities, rather than hope that they may somehow mould themselves seamlessly and benignly to us" (Johnston et al., 2007). Recent research in ecosystem ecology has brought significant new knowledge about the "general tendencies of ecosystem properties and processes" both in their normal state of growth and development and when under disturbance (Nielsen et al., 2020). According to the book entitled *A New Ecology Systems Perspective*, ecosystems have "directionality" (Jørgensen and Mejer, 1977), as they always develop toward (in the direction of) increasing the amount of exergy stored in the system, exergy being the net amount of total energy that

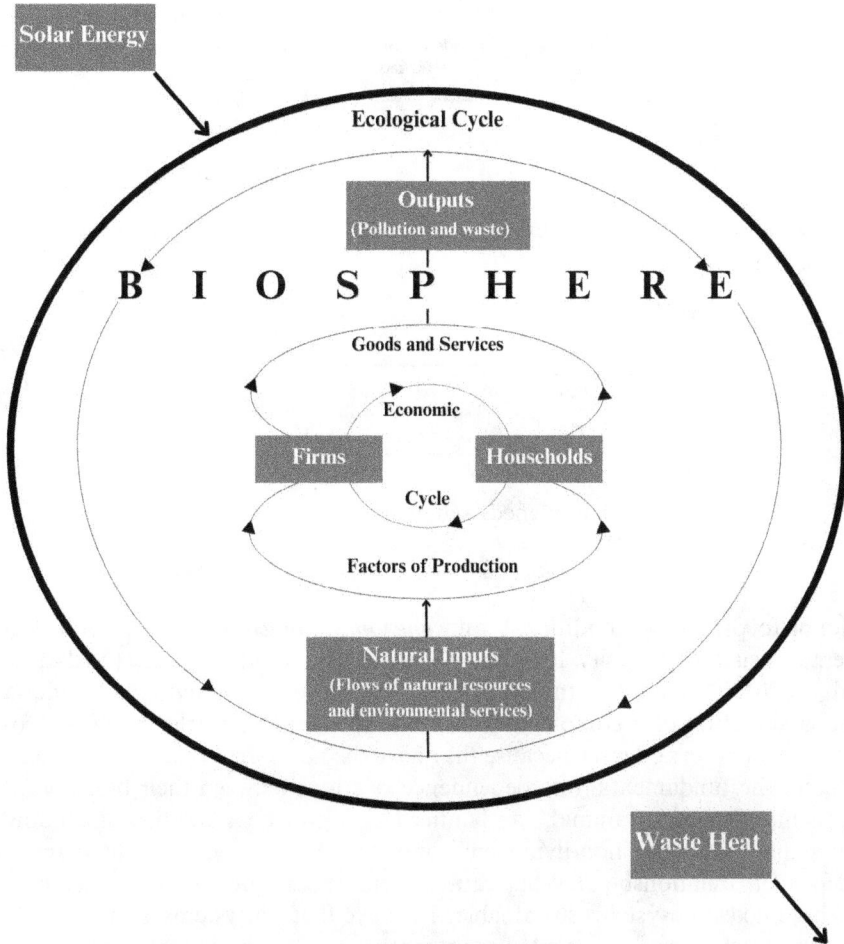

Figure 3.1 The economy embedded in biosphere
Source: Daly and Farley, 2011.

can be converted directly into work and can help the system grow and develop. On the other hand, ecosystems are open systems subject to the laws of thermodynamics. They are also characterized by ontic openness "due to the complex web of life constantly combining, interacting, and rearranging in the natural world to form novel patterns" (Nielsen et al., 2020), meaning that we should expect unpredictability, uncertainty, emergence, and self-organization both in the structure and the behaviour of ecosystems, as if they had a life of their own that we hardly understand or control. The book identifies nine principles which characterize any ecosystem. These principles are organized as *material constraints* (imposed by the laws of thermodynamics and the periodic table of elements), *ontological properties* (ecosystems are self-organizing, co-evolve, and

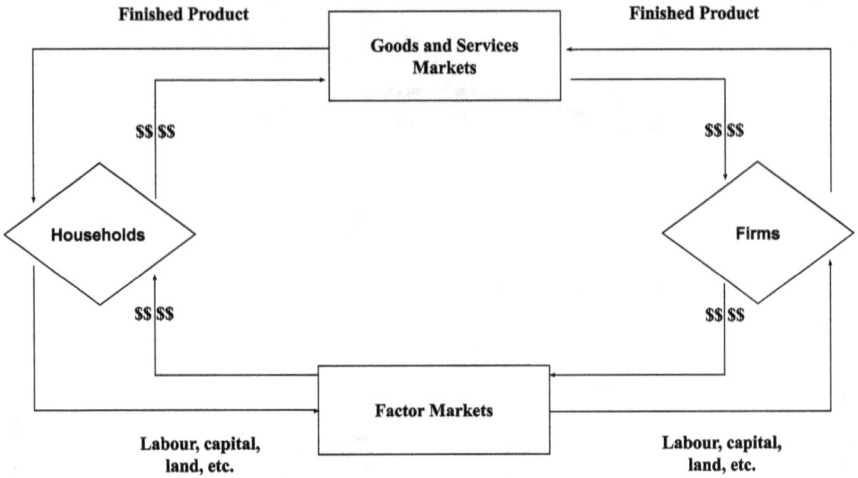

Figure 3.2 The circular flow of goods and money
Source: Internet: ThoughtCo.

adapt to prevailing conditions), and *phenomenological properties* (observed features such as diversity, hierarchies, networks, and information) (Nielsen et al., 2020). This view of the ecosystem's structure and functioning paints a holistic picture of a complex, unitary but probabilistic world in which ecosystems survive and grow because they have the necessary conditions, but also due to the fundamental interdependence relations between their biotic components – plant, animal, and microorganism communities, including humans – and the non-living environment. What underlines these interdependent relationships? What causes them and secures their continuance? What makes ecosystems sustainable? I believe that ecosystems' sustainability is due to the special role played in ecosystems by the biotic components, which are alive and able to participate in the Universe's life cycle on account of some *real universal life force* inherently existing in them and producing sustaining effects in the whole ecosystem. Thus, even the inanimate matter is enfolded in the life process, as, for instance, when a molecule of carbon dioxide crosses a cell boundary and becomes "alive" in a leaf. To explain this, we must assume that "life itself" belongs "in some sense to a totality, including plant and environment" (Bohm, 1980). We can understand sustainability as a recurrent process of "enfolding" and "unfolding" of ecosystem components, which keeps the ecosystem in continuous existence and emergence. Bohm illustrates this process using a forest "constituted of trees that are continually dying and being replaced by new ones. If it is considered on a long time-scale, this forest may be regarded likewise as a continuously existent but slowly-changing entity" (Bohm, 1980). These processes are not always visible or easily understandable due to the complexity of their intricacies both in structure and function. Rather, scientists stumble upon them and are always surprised when they do, as they tend to surpass human comprehension and require openness to conjectures. For instance, Newton discovered the law of universal gravity in "a sudden flash of insight" when he "saw that as the apple falls so does the moon, and so indeed do all objects" – "seen as falling toward various centres (e.g. the earth, the sun, the planets, etc.)" (Bohm, 1980). Newton's insight meant his understanding of the Universe's wholeness, with no longer a division between Heaven and Earth.

Likewise, in the 1940s, as the physicist Erwin Schrödinger was trying to understand the paradox of what explains the miraculous ability of living organisms to delay their decay into thermodynamic equilibrium (death), if death and decay is the fate of all things according to the second law of thermodynamics, his research into the physical aspects of the living cell led him to an important discovery which paved the way to the science of genetics. Schrödinger defined life as "orderly and lawful behaviour of matter, not based exclusively on its tendency to go over from order to disorder, but based partly on existing order that is kept up" (Schrödinger, 1944). He gave two explanations for what is "keeping up" order in living beings. First, a living organism, be it a microorganism, plant, animal, or human, is an open system exchanging energy, mass, and information with the environment, and thus being able to grow and postpone reaching the state of death (thermodynamic equilibrium) by "feeding upon negative entropy, attracting, as it were, a stream of negative entropy upon itself, to compensate the entropy increase it produces by living and thus to maintain itself on a stationary and fairly low entropy level" (Schrödinger, 1944). This process can be exemplified by a human being eating an apple (thus introducing negative, low entropy in the body), or a human being smoking (introducing high entropy in the body). Second, digging deeper into microbiology, Schrödinger explained that "in order to understand life – the stability and permanence of the genetic material – classical physics was inadequate, and one had to go to the very basis of quantum theory" (Symonds, 1986). He suggested that hereditary properties of the genes are encoded in an "aperiodic crystal" or solid, a "tiny speck of material, the nucleus of the fertilized egg", which "contains an elaborate code-script involving all the future development of the organism" (Schrödinger, 1944). Schrödinger's insight that the gene could be considered an aperiodic crystal containing the code-script of life inspired research in microbiology that led to discovery of DNA as the genetic material, and discovery of its double-helix structure in 1953 (Sigmund, 2019). However, Schrödinger was honest to acknowledge that "living matter, while not eluding the 'laws of physics' as established up to date, is likely to involve 'other laws of physics' hitherto unknown" (Schrödinger, 1944). The discovery of DNA, which explains how living cells live, process, replicate, and transmit hereditary information stored in the double-helix code, created another enigma: what is the origin of the complex, specified information stored in the gene molecule? Could the digital code stored in DNA show intelligent thinking and planning of the molecular structures and processes that synthesize the proteins sustaining the living cell? Scientists were puzzled and answered these questions according to their worldviews. The Neo-Darwinists marvelled at the "digital code" stored in the DNA (Hood and Galas, 2003), or discussed the fact that "the machine code of the genes is uncannily computer-like" (Dawkins, 1995), but did not pay attention to the "philosophical" question of who the author of the precise assembly instructions in the DNA for building the protein molecules that sustain life in the cell might be. At the same time, a series of scientists, mostly multidisciplinary, started discussing these deeper questions, such as the origin of life which in their view required an "intelligent cause" (Thaxton et al., 1984), or the criteria by which intelligent causes can be identified in the functioning of our Universe. The mathematician and philosopher William Dembski (1998) identified two such criteria which "invariably" denote prior intelligent activity: *complexity* (or small probability) of events or systems, and *specification* (as a special kind of recognizable pattern or functional requirement). For him, the origin of the complex, specified information in the DNA and the proteins needed for building and sustaining the living cell can only be attributed to an intelligent cause. Other teleological arguments were brought by biochemist Michael Behe, who saw designed "irreducible complexity" in mechanisms sustaining life, like the blood-clotting mechanism (Behe, 1996), philosopher of science Stephen Meyer in his book *Signature in the Cell* (2009), and philosopher Thomas Nagel (2012) who discussed why he believed that "rational intelligibility is at the root of the natural order".

I agree that finely crafted and precisely functioning mechanisms like the DNA molecule suggest design and intentionality to support life at the molecular level. This can be obvious to molecular biochemists, geneticists, and medical doctors exploring life through microscopes. Brazilian chemist Marcos Eberlin, for instance, refers to the way any living cell, life's most basic unit, is protected by a carefully engineered "double-layer membrane – flexible, stable and resistant – [...] one that would promptly and efficiently protect the cell from the devastating O_2 permeation, can remain stable in aqueous acid media and ably handle fluctuations in temperature and pH" (Eberlin, 2019). Biochemist and medical doctor Michael Denton in his book *The Miracle of Man*, describes the human brain as possibly "the most complex functional organization of matter possible" (Denton, 2022):

> Each cubic millimeter of human cerebral cortex (the thinking part of the brain) contains 15,000 neurons, 15,000 glial cells (which provide support for the neurons), some four kilometers of axonal wiring, 500 meters of dendrites, and 400 million synaptic connections (the highest recorded density of synapses in any mammalian brain).
>
> (Denton, 2022)

There are numerous such fine-tuned mechanisms in the living world which speak about intelligent behaviours in nature meant to protect life. For example, trees transfer sap to the roots as winter approaches to protect their trunks from freezing during frigid winters. Some molecular botanists believe that tree roots act like brain-structures for helping trees live and thrive (Baluška et al.,

2007). Migratory birds "know" each fall that it is time to migrate to warmer places and to return in spring to their native places. One bird species, the Eurasian reed warbler (*Acrocephalus scirpaceus*), uses the Earth's magnetic inclination to precisely execute their long migratory trips (Wynn et al., 2022). These mechanisms do not only denote intelligent activity, but they also reveal ingenious solutions in nature that humans have "mimicked" throughout history to build their advanced civilization upon, from the discovery of the secrets of the atom to quantum cryptography and to medical solutions to free radicals in microbiology. They clearly show that ecological sustainability or perpetuation of life on Earth is the purpose intended in the Universe, and they indicate the way of achieving it, through intelligent action that carefully observes the existing boundaries and chooses only interventions that do not damage but tend to "preserve the integrity, stability and beauty of the biotic community" (Leopold, 1949) at the heart of sustainability. Humans have a special place in the biotic community, by being endowed with reason, conscience, and moral sentiments. They can discover more mysteries of the Universe which can make life on Earth not only feasible but enjoyable. The experiments of the 2022 Nobel Prize winners in physics – Alain Aspect, John F. Clauser, and Anton Zeilinger – have confirmed that quantum mechanics' strange description of reality given by Schrödinger's wave function is correct, and that the quantum entanglement phenomena can provide "the tools to use distant, but still entangled, photons, such as Bell pairs, as a quantum resource" in "the rapidly developing field of quantum information science." (Nobel Prize in Physics, 2022). We can explain the source of these sometimes hard-to-understand "quantum resources" either by assuming that they have evolved through random mutations, or by choosing to believe that there is a cosmic Mind that designed life mechanisms at the molecular level, and also sustains life in the quantum and the macro Universe. I choose this latter explanation, based on the empirical, actual, and real aspects of life documented in this book, as this explanation puts me in my place as a rational and moral being, giving me the reason to live and act consistently in everyday life. My eyes, my mind, and science tell me that the Universe and the planet I live on do not just appear to be orchestrated but are indeed carefully and intelligently crafted in a way that I can mostly understand, and through this, they make me feel a welcomed and necessary part of the living word. This world was conceived to function and persist as a whole through the interaction of a multitude of living beings, each doing their part, and as a living being, I feel compelled to fulfil my sacred part by participating responsibly in the planetary life cycle, a cycle which I do not control, but which I can impact through my decisions. It was Aristotle in his *Physics* and in his *Metaphysics* who told us that as terrestrial beings we have a purpose and we need to act in order to achieve that purpose (Sachs and Aristotle, 1995). Aristotle saw the Universe as a whole and every being in the Universe as a whole aiming to achieve its own wholeness and contribute to the wholeness of the

Universe. That wholeness provides a stable condition for the flourishing of plants, animals, and of humans in lives and acts that come to completion in their own ways. In Aristotle's view, "life is not a strange by product of things, but the source of things, and the non-living side of nature has being in a way strictly analogous to life: as an organized whole that maintains itself by continual activity" (Sacks and Aristotle, 1995).

The idea of undivided wholeness of things also transpires in the theory of relativity and in the quantum theory (Bohm, 1980), challenging the notion that reality exists at random and functions as an unthinking machine. In 1944, German physicist Max Planck said:

> There is no matter as such. All matter originates and exists only by virtue of a force which brings the particle of an atom to vibration and holds this most minute solar system of the atom together. We must assume behind this force is the existence of a conscious and intelligent Mind. This Mind is the matrix of all matter.
>
> (Planck, 1944)

In order to become sustainable, humanity needs to see sustainability in nature as a non-random way of organizing the whole Universe in order to maintain itself and to protect the life of each living creature. We also need to change our attitude toward quantum mechanics, which is currently mostly celebrated for its weirdness and for the fact that it does not make sense (Berlinski, 2023). This needs to change. As an economist interested in the world of sustainability, I want to understand what the quarks really are, what the quantum field is, and what the newly discovered "quantum resources" are. I believe we do not teach enough quantum mechanics to our students, and we should, including the "spooky aspects" that science cannot explain. Doing more research in quantum mechanics, we might discover, like Erwin Schrödinger, hints about the author of the life-sustaining system. In 1944, in his book *What Is Life?*, Schrödinger noted that the chromosome fibres in a living organism are not human-made, but "the finest masterpiece ever achieved along the lines of the Lord's quantum mechanics" (Schrödinger, 1944).

In conclusion, sustainability exists objectively in nature, embedded in numerous purposeful beings and in minutiously crafted life-sustaining mechanisms; it can be known both in its common-sense aspects and its "spooky" quantum-mechanics aspects, which suggest that a super-intelligent Being transcending space, time, and the speed of light is running the complex, dynamic life-sustaining system. The intricate and exquisite make-up of our life-sustaining Universe, which is freely available to all living things, makes some demands on human beings, as the only living creatures endowed with reason, conscience, and moral sentiments. The way human beings relate to sustainability in nature is the topic of the next chapter.

References

Baluška, F. *et al.* (2007) "Neurobiological View of Plants and Their Body Plan". In *Communication in Plants*, edited by F. Baluška, S. Mancuso and D. Volkmann. New York: Springer.

Barnes, L.A. (2020) "A Reasonable Little Question: A Formulation of the Fine-Tuning Argument." *Ergo*, 6(42): 1220–1257.

Barrow, J.D. and Tipler, F.J. (1986) *The Anthropic Cosmological Principle*. Oxford: Oxford University Press.

Behe, M.J. (1996) *Darwin's Black Box: The Biochemical Challenge to Evolution*. New York: The Free Press.

Berlinski, D. (2023) *Science After Babel*. Seattle: Discovery Institute Press.

Bhaskar, R. (1975) *A Realist Theory of Science*. Leeds, UK: Leeds Books.

Bohm, D. (1980) *Wholeness and the Implicate Order*. London and New York: Routledge.

The Born-Einstein letters: correspondence between Albert Einstein and Max and Hedwig Born from 1916–1955, with commentaries by Max Born (1971). Macmillan.

Britannica, The Editors of Encyclopaedia (n.d.) "Schrödinger equation." Encyclopedia Britannica. www.britannica.com/science/Schrodinger-equation.

Britannica, The Editors of Encyclopaedia (n.d.) "Naturalism." Encyclopedia Britannica. www.britannica.com/topic/naturalism-philosophy.

Buranyi, S. (2022) "Do we need a new theory of evolution?" *The Guardian*, June 28. www.theguardian.com/science/2022/jun/28/do-we-need-a-new-theory-of-evolution.

Burbidge, E.M. (1957) "Synthesis of the Elements in Stars." *Reviews of Modern Physics*, 29: 547–650. doi:doi:10.1103/RevModPhys.29.547.

Clausius, R. (1865) *The Mechanical Theory of Heat – with its Applications to the Steam Engine and to Physical Properties of Bodies*. London: John Van Voorst.

Cohen, I. (1978) *Isaac Newton's Papers and Letters on Natural Philosophy and Related Documents*. Cambridge, MA: Harvard University Press.

Collins, C.B. and Hawking, S.W. (1973) "Why Is the Universe Isotropic?" *Astrophysical Journal*, 180: 317–334. http://dx.doi.org/10.1086/151965.

Comte, A. (1858) *The Positive Philosophy of Auguste Comte*, translated by H. Martineau. New York: Kalvin Blanchard.

Craig, W.L. (2011) "Graham Oppy on the Kalam Cosmological Argument." *International Philosophical Quarterly*, 51: 303–330.

Daly, H. and Farley, J. (2011) *Ecological Economics: Principles and Applications*. Washington, DC: Island Press.

Darwin , C. (1860) *On the Origin of Species by Means of Natural Selection, or the Preservation of Favoured Races in the Struggle for Life*. New York: D. Appleton and Company.

Darwin, C. (1871) *The Descent of Man, and Selection in Relation to Sex*. London: John Murray.

Dawkins, R. (1995) *River out of Eden: A Darwinian View of Life*. New York: Basic Books.

Dembski, W.A. (1998) *The Design Inference: Eliminating Chance Through Small Probabilities*. Cambridge: Cambridge University Press.

Denton, M. (1985) *Evolution: A Theory in Crisis*. London: Adler and Adler.

Denton, M. (2017) *The Wonder of Water: Water's Profound Fitness for Life on Earth and Mankind*. Seattle, WA: Discovery Institute Press.

Denton, M. (2022) *The Miracle of Man. The Fine Tuning of Nature for Human Existence*. Seattle, WA: Discovery Institute Press.

de Vries, B.J.M. (2012) *Sustainability Science*. Cambridge: Cambridge University Press.

Dodd, M.S., Papineau, D., Grenne, T. et al. (2017) "Evidence for early life in Earth's oldest hydrothermal vent precipitates." *Nature*, 543 (7643): 60–64. https://doi.org/10.1038/nature21377.

Dyson, F. (2001) *Disturbing the Universe*. New York: Basic Books.

Eberlin, M. (2019) *Foresight: How the Chemistry of Life Reveals Planning and Purpose*. Seattle, WA: Discovery Institute Press.ray

Egbert, R.G. and Klavins, E. (2012) "Fine-tuning gene networks using simple sequence repeats." *Proceedings of the National Academy of Sciences*, 109(42): 16817–16822. www.pnas.org/cgi/doi/10.1073/pnas.1205693109.

Einstein, A., Podolsky, B., and Rosen, N. (1935) "Can Quantum-Mechanical Description of Physical Reality be Considered Complete?" *Physical Review*, 47(10): 777.

Einstein, A. (1993) "Physics and Reality". In *Out of My Later Years: The Scientist, Philosopher, and Man Portrayed Through his Own Words* (1993). New York: Wings Books. (Original work published 1936)

Eliade, M. (1978) *A history of religious ideas*, Vol. 1. Chicago: University of Chicago Press.

Fiscus, D.A. and Fath, B.D. (2019) *Foundations for Sustainability A Coherent Framework of Life-Environment Relations*, London: Academic Press.

Galilei, G. (1960) *The Assayer, in The Controversy of the Comets of 1618*, edited by S. Drake. Philadelphia: University of Pennsylvania Press.

Georgescu-Roegen, N. (1971) *The Entropy Law and the Economic Process*. Cambridge, MA: Harvard University Press.

Giberson, K.W. (2012) *The Wonder of the Universe: Hints of God in Our Fine-Tuned World*. Downers Grove, IL: InterVarsity Press.

Gunderson, L.H. and Holling, C.S. (2009) *Panarchy Understanding Transformations in Human and Natural Systems*. Washington, DC: Island Press.

Henderson, L.J. (1913) *The Fitness of the Environment. An Inquiry into the Biological Significance of the Properties of Matter*. New York: The Macmillan Company.

Hood, L. and Galas, D. (2003) "The Digital Code of DNA." *Nature*, 421: 444–448.

Hoyle, F. (1982) "The Universe: Past and Present Reflections." *Annual Review of Astronomy and Astrophysics*, 20: 1–35. www.annualreviews.org/doi/pdf/10.1146/annurev.aa.20.090182.000245.

Hume, D. (1777) "An Enquiry Concerning Human Understanding." In *Enquiries Concerning Human Understanding and Concerning the Principles of Morals*, edited by L.A. Selby-Bigge (1902). Oxford: Clarendon.

Huxley, T.H. (1871) *Lay sermons, addresses and reviews*. New York: Appleton.

Iacoboni, S.J. (2022) *Telos: The Scientific Basis for a Life of Purpose*. Savage, MI: BroadStreet Publishing.

Janet, P. (1878) *Final causes*, translated by W. Affleck. Edinburgh: Clark.

Johnston, P., Everard, M., Santillo, D., and Robèrt, K-H. (2007) "Reclaiming the Definition of Sustainability." *Environmental Science and Pollution Research*, 14(1): 60–66.

Jørgensen, S.E. and Mejer, H.E. (1977) "Ecological buffer capacity." *Ecological Modelling*, 3: 39–61.

Kaiser, C. (1991) *Creation and the History of Science*. Grand Rapids, MI: Eerdmans.

Kant, I. (2012) *Immanuel Kant: Natural Science. The Cambridge Edition of the Works of Immanuel Kant*, edited by E. Watkins. Cambridge: Cambridge University Press. (Original work published 1755)

Kant, I. (2004) *Critique of Practical Reason*. Mineola, NY: Dover Publications.

Kant, I. (1968) *Critique of Pure Reason*, edited by N. Kemp Smith. London: Macmillan. (Published as *Kritik der reinen Vernunft*, 2nd ed., 1787)

Kant, I. (1952) *The Critique of Judgement*, translated by J.C. Meredith, edited by R. M. Hutchins. Chicago, London, Toronto: Encyclopaedia Britannica. (Published as *Kritik der Urteilskraft*, 1790)

Laland, K., Uller, T., Feldman, M.*et al.* (2014) "Does evolutionary theory need a rethink?" *Nature*, 514: 161–164. https://doi.org/10.1038/514161a.

Lenton, T.M., Rockström, J., Gaffney, O., Rahmstorf, S., Richardson, K., Steffen, W., and Schellnhuber, H.J. (2019) "Climate tipping points — too risky to bet against. The growing threat of abrupt and irreversible climate changes must compel political and economic action on emissions." *Nature*, 575: 592–595. https://doi.org/10.1038/d41586-019-03595-0.

Leopold, A. (1949) *A Sand County Almanac and Sketches Here and There*. Oxford: Oxford University Press.

Lewis, G. F. and Barnes. L.A. (2016) *A Fortunate Universe: Life in a Finely Tuned Cosmos*. Cambridge: Cambridge University Press.

Loeb, A. (2020) "Living with Scientific Uncertainty." *Scientific American*, July 15, 2020. www.scientificamerican.com/article/living-with-scientific-uncertainty.

MacIntyre, A. (1981) *After Virtue: A Study in Moral Theory*. Notre Dame, IN: University of Notre Dame Press.

Maxwell, C. (1873) "Address of the British Association" In *Scientific Papers* (Vol. 2) (1890). Cambridge: Cambridge University Press.

Meyer, S. (2009) *Signature in the Cell*. New York: HarperOne.

Meadows, D. (2008) *Thinking in Systems: A Primer by Donella Meadows and Diana Wright*, Vermont: Chelsea Green Publishing.

Millennium Ecosystem Assessment (2005) *Living Beyond Our Means: Natural Assets and Human Well-being*. Washington, DC: Island Press.

Nagel, T. (2012) *Mind and Cosmos: Why the Materialist Neo-Darwinian Conception of Nature is Almost Certainly False*. Oxford University Press.

New Hubble (2019) "Constant Measurement Adds to Mystery of Universe Expansion Rate." https://hubblesite.org/contents/news-releases/2019/news-2019-28.

Newton, I. (1962) *Sir Isaac Newton's Mathematical Principles of Natural Philosophy and His System of the World*, edited by F. Cajori, translated by A. Motte. Berkeley: University of California Press. (Original work published 1687)

Nielsen, S.N., Fath, B.D., Bastianoni, S., Marques, J.C., Muller, F., Patten, B. C., Ulanowicz, R.E., Jorgensen, S.E., and Tiezzi, E. (2020) *A New Ecology Systems Perspective*, 2nd ed. Amsterdam, Cambridge, MA: Elsevier.

Nobel Prize in Physics (2022) "Scientific Background on the Nobel Prize in Physics 2022: 'For Experiments with Entangled Photons, Establishing the Violation of Bell Inequalities and Pioneering Quantum Information Science'." The Royal Swedish Academy of Sciences. www.nobelprize.org/uploads/2022/10/advanced-physicsprize2022.pdf.

Odum, E.P. (1977) "The emergence of ecology as a new integrative discipline." *Science*, 195(4284): 1289–1293.

Patterson, C.C. (1956) "Age of meteorites and the Earth." *Geochimica and Cosmochimica Acta*, 10(4): 230–237.

Perlmutter, S., Aldering, G., Goldhaber, R., Knop, A., Nugent, P., and Castro, P.G. (1999) "The Supernova Cosmology Project. Measurements of Ω and Λ from 42 High-Redshift Supernovae." *The Astrophysical Journal*, 517(2): 565–586. https://doi.org/10.1086/307221.

Planck, M. (1944) "Das Wesen der Materie" (The Nature of Matter). Speech at Florence, Italy.

Planck (2015) "Planck science highlights." ESA (The European Space Agency) https://www.esa.int/Science_Exploration/Space_Science/Planck/Planck_science_highlights Accessed December 1st, 2023.

Rees, M. (2000) *Just Six Numbers: the Deep Forces that Shape the Universe.* New York: Basic Books.

Richter, D., Grün, R., Joannes-Boyau, R., Steele, T.E., Amani, F., Rué, M., Fernandes, P., Raynal, J-P., Geraads, D., Ben-Ncer, A., Hublin, J-J., and McPherro, S. P. (2017) "The age of the hominin fossils from Jebel Irhoud, Morocco, and the origins of the Middle Stone Age." *Nature*, 546: 293–296.

Riess, A.G., Filippenko, A.V., Challis, P.*et al.* (1998) "Observational Evidence from Supernovae for an Accelerating Universe and a Cosmological Constant." *The Astronomical Journal*, 116(3): 1009–1038. https://doi.org/10.1086/300499.

Robbins, J. (2012) "Why Trees Matter." *The New York Times*, April 11, 2012. www.nytimes.com/2012/04/12/opinion/why-trees-matter.html?_r=1&.

Rockström, J., Steffen, W., Noone, K.*et al.* (2009) "A safe operating space for humanity." *Nature*, 461: 472–475. https://doi.org/10.1038/461472a.

Rosen, R. (1991) *Life Itself: A Comprehensive Inquiry into the Nature, Origin, and Fabrication of Life.* New York: Columbia University Press.

Sachs, J. and Aristotle (1995) *Aristotle's Physics: A Guided Study.* New Brunswick, NJ: Rutgers University Press.

Schrödinger, E. (1935) "Discussion of probability relations between separated system." *Mathematical Proceedings of the Cambridge Philosophical Society*, 31(4): 555–563.

Schrödinger, E. (1944) *What Is Life?* Cambridge: Cambridge University Press.

Sigmund, K. (2019) "The physicist and the dawn of the double helix Erwin Schrödinger's prescient musings on molecular biology turn 75." *Science*, 366(6461). https://doi.org/10.1126/science.aaz4846.

Sinclair, D.A. with LaPlante, M.D. (2019) *Lifespan. Why We Age, and Why We Don't Have To.* New York: Atria Books.

Singh, S. (2005) *Big Bang: The Origin of the Universe.* New York: HarperCollins.

Spencer, H. (1864) *The Principles of Biology*, Vol. 1. London: Williams and Norgate.

Symonds, N. 1986 "What Is Life? Schrödinger's Influence on Biology." *The Quarterly Review of Biology*, 61(2): 221–226.

Tegmark, M., Aguirre, A., Rees, M.J., and Wilczek, F. (2006) "Dimensionless Constants, Cosmology, and Other Dark Matters." *Physical Review D*, 73(2): 023505. https://doi.org/10.1103/PhysRevD.73.023505.

Thaxton, C., Bradley, W.L., and Olsen, R.L. (1984) *The Mystery of Life's Origin: Reassessing Current Theories.* New York: Philosophical Library.

Thorvaldsen, S. and Hössjer, O. (2020) "Using statistical methods to model the fine-tuning of molecular machines and systems." *Journal of Theoretical Biology*, 501: 110352. https://doi.org/10.1016/j.jtbi.2020.110352.

Vitousek, P.M., Mooney, H.A., Lubchenco, J., and Melillo, J.M. (1997) "Human Domination of Earth's Ecosystems." *Science*, 277: 494–499.

Wackernagel, M. and Rees, W.E. (1995) *Our Ecological Footprint: Reducing Human Impact on the Earth.* New Society Publishers.

Wagner, A. (2014) *Arrival of the Fittest: How Nature Innovates.* New York: Current.

Weinberg, S. (1989) "The Cosmological Constant Problem." *Reviews of Modern Physics*, 61(1): 1–23. https://doi.org/10.1103/RevModPhys.61.1.

Wells, J. (2017) *Zombie Science: More Icons of Evolution.* Seattle, WA: Discovery Institute Press.

Woese, C.R. (2004) "A New Biology for a New Century." *Microbiology and Molecular Biology Reviews*, 68(2): 173–186.

Wohlleben, P. (2015) *The Hidden Life of Trees. What They Feel, How They Communicate.* Vancouver and Berkeley: Greystone Books.

Wohlleben, P. (2021) *The heartbeat of trees. Embracing our ancient bond with trees and nature.* Vancouver and Berkeley: Greystone Books.

Wynn, J., Padget, O., Mouritsen, H., Morford, J., Jaggers, P., and Guilford, T. (2022) "Magnetic stop signs signal a European songbird's arrival at the breeding site after migration." *Science*, 375(6579): 446–449.

4 Restoring sustainability as objective social reality

We can define social reality as an objective system made up of human beings, aggregated in families, communities, and nations, and their needs as objectively defined (Gough, 2017), plus human-made artefacts that can be material (e.g. buildings, money, technologies, means of transportation) or immaterial and abstract entities (e.g. institutions, laws, debt, inflation, class structures, thoughts, poetry, faith, and gender relations), all embedded in the Earth's biosphere. The fact that societies are made of people who behave towards one another in certain ways is a truism that few would deny (Ferraris and Torrengo, 2014). It is also a truism that people interact with nature, as they have material bodies that "cannot exist without intact ecosystems, fertile soils, drinkable water, and a reasonably intact global climate" (Ekardt, 2020). At the same time, humans live according to the laws of nature which are incorporated, as axioms, into their daily lives. Then we can safely say that a society is not a static entity but a dynamic "ensemble of structures, practices and conventions which individuals reproduce or transform, but which would not exist unless they did so" (Bhaskar, 1979). That humans have a limited lifespan of up to 120 years is also a truism, confirmed by the history of humanity, as "most of us, 95 percent to be precise, are dead before 100" (Sinclair, 2019). While human mortality can be explained by the law of entropy in physics, Schrödinger noted as early as 1944 the existence of metabolism as the "marvellous faculty of any living organism" by which it is able to delay the decay into thermodynamic equilibrium (death) by "feeding on a stream of negative entropy to compensate the entropy increase it produces by living", and thus maintaining itself on a stationary and fairly low entropy level (Schrödinger, 1967). Research in biology has confirmed that highly fine-tuned metabolic processes take place in the living cell, where the code-like information in the DNA molecule guides the building of proteins and protein machines performing essential life-maintaining functions (Thorvaldsen and Hössjer, 2020). While human lifespan is limited at the level of individuals, "quasi-immortal entities – such as a nation and especially mankind" may exist indefinitely, as they are not "subject to a mortality table" (Georgescu-Roegen, 1979). History shows that civilizations may collapse, as the Roman empire, and the Mayan or the Easter Island civilizations did, but societies survive or are sustainable

DOI: 10.4324/9781003307587-5

even if under changed forms and with reduced populations (Tainter, 1988). Nature and its provision of life-sustaining conditions for living beings is thus objectively part of the social system, however, not in its pristine, integral, primordial form, as over millennia humans have used their ability to modify their environment, and very few untouched areas still exist on Earth. Consequently, "most of life for most people takes place on those anthropogenic biomes that are a hybrid tapestry of nature and culture" (Rolston, 2017). Humans' impact on nature, in the geological epoch which some call the Anthropocene (Adeney, 2022), is so prevalent that some authors have started to assert that "Nature no longer runs the Earth. We do." We are "the God species" (Lynas, 2011) or "Homo Deus" (Harari, 2015). Some other authors believe that humans' main task is not to run the Earth but to "heal the planet" and "learn again to live as part of nature" (Washington, 2015). Still others believe that in order to preserve our life-friendly planet for the future, humans need to be less concerned about sustaining a managed planet, rebuilt for human benefit, and focus more on caring for it and sustaining the biosphere we have inherited (Princen, 2010; Rolston, 2017). These are conflicting views about *Homo sapiens'* place and role in the world which warrant an investigation about humans' concern for sustainability on the built planet, with a focus on their contribution or lack thereof to secure future generations and their survival.

We start our analysis from the premise deduced from our definition of sustainability that human life is valuable both now and in the future, and from the stark fact that the last 100 years have been the bloodiest in human history, much darker than the so-called Dark Ages (450–1450 AD). Approximately 231 million men, women, and children died violently in wars and conflicts in the twentieth century (Leitenberg, 2006), including the horrors of the Holocaust, the Gulag, the atrocities of the Chinese civil war and Communist and Cultural revolutions, plus "various massacres, sub-continental famines, squalid civil insurrections, blood-lettings, throat-slittings, death squads, theological infamies, and suicide bombings taking place from Latin America to East Timor" (Berlinski, 2019). Why is that? Why are most societies in the current predicament where life on Earth has become violent and unsustainable? Why have the social developments predicted since the nineteenth century by the "philosophers of progress", such as Marx, Spencer, or Dewey, not actually taken place? What went wrong with liberal societies' claim "that human social development is inevitably melioristic" (Barrow and Tipler, 1986)? Will it ever become the accepted norm that our socio-economic systems should be organized in sustainable ways which exclude violence to nature and other human beings, or are we doomed to continue destroying our life-sustaining systems and each other? What could motivate humans to choose sustainable lifestyles? What could bring social change for sustainability? This chapter will attempt to answer these questions using insights from the philosophy of critical realism. According to critical realism, reality exists objectively and is stratified in three distinct domains: experiential reality, actual reality, and real (transcendental) reality (Bhaskar, 1975). In the same way, each individual human being is conceptualized as a self, existing and acting at three differentiated levels – the ego, the actual (embodied) self, and the transcendentally real self (Bhaskar, 2012). All humans are social beings involved in social acts taking place at four specific levels, namely of "our material transactions with nature, our social interactions with others, the level of social structure and the level of the stratification of our own embodied personalities" (Bhaskar, 2012).

I argue that social change towards sustainability is possible at the individual level, given that humans are intentional and reflexive social beings, able to choose between good and evil, as "the line separating good and evil passes not through states, nor between classes, nor between political parties either – but right through every human heart – and through all human hearts" (Solzhenitsyn, 2020). Every human is able to conceive sustainability rationally, and to aim to achieve it as a human aspiration, but not based on science alone, as science can only explain our material world, not our aspirations and deep longings. We can become sustainable beings when we accept that we are not only material beings but also spiritual beings with ultimate goals whose origin is "from another source" (Einstein, 1954). However, our ability to act like free and strong, "whole" agents of change towards sustainability has been impeded in the last three centuries by the misguided view that human nature does not exist intrinsically, and by the widespread confusion – brought about by the modernity thinking – about how the world really functions and what sustainability is. I will discuss first human nature, defined as the view that humans have some essential, defining properties "not accidentally but necessarily" (Berlinski, 2019).

Human nature

It is useful to start our analysis of human nature with Aristotle's thoughtful explanation of what a "being" is in its essence, meaning being "as such" or "in its own right", as described in Aristotle's *Physics and Metaphysics* (Sachs, 1995). For Aristotle, "being" consists of "thinghood" defined as "the essence (*ousia*) of a thing" or "what it keeps on being in order to be at all (*to ti n einai*), and must be a being at work (*energeia*) so that it may achieve and sustain its being at work staying itself (*entelecheia*)" (Sachs, 1995). In other words, the being is a dynamic "whole that maintains itself by its own activity" and is also an undivided whole that belongs to another undivided whole, the Universe. "What Aristotle in fact means is that every natural being is a whole, and every natural activity leads to or sustains that wholeness." (Sachs, 1995). We understand that Aristotle's being is a living being when he notes that the being is "the sort of being that belongs only to animals, plants, and the cosmos as a whole" (Sachs, 1995). In addition, Aristotle shows, referring to his famous four causes of things, that "Every being consists of material and form, that is, of an inner striving spilling over into an outward activity. Potency and being at work are the ways of being of material and form"

(Sachs, 1995). That means that a flower bud is just matter and potential until "the inner striving spills over into" the form of a beautiful flower. Why does a beautiful flower exist at all? To accomplish its purpose, or its "final cause" defined by Aristotle as "that for the sake of which" something does what it does or is what it is (Sachs, 1995). Does a flower exist for human enjoyment? Maybe, but we know for sure that a flower exists for the sake of bees and other insects visiting it to collect pollen and secure pollination "in a cycle of ever renewed wholeness" of the Universe which "provides a stable condition for the flourishing of plants and of humans in lives and acts that come to completion in their own ways" (Sachs, 1995). According to Aristotle, human "beings" exist as highly organized matter, form, and energy and have a "telos", namely to act and achieve "wholeness" in their complete life, a life well lived, as an intrinsic part of the "wholeness" of the sustaining Universe.

Homo sapiens in fact

While being part of the animal world biologically, humans have certain abilities which set them apart from animals and give them a special "calling as the only morally responsible beings in the world" (Polanyi, 1958/1976). Polanyi argued that morality, defined as the intrinsic capacity to distinguish between good and evil, right and wrong, and the moral responsibility that comes with it, is what separates humans from animals. This view was also held by ancient philosophers (Aristotle, Augustine) and medieval theologians (Aquinas); however, it started to change during the Enlightenment, when commentators began nominating language (Kagan, 2004), or humans' advanced intellectual abilities (reason), self-awareness, and ability to think abstract thoughts (Ayala, 2010) as features holding humans' defining characteristic. Indeed, humans are endowed with "mental incandescence" (Rolston, 1989), or mental faculties that enable them to think, reason and comprehend, investigate, make decisions, and choose courses of action. Nagel considers humans' faculty of reason "an irreducible faculty", which has enabled them to "transcend the perspective of the immediate life-world given to us by our senses and instincts, and to explore the larger objective reality of nature and value" (Nagel, 2012). Human ingenuity and humans' ability to reflexively monitor their courses of action have led during the last century to numerous discoveries which have allowed humans to build the current scientifically and technologically advanced civilization. However, studies show that human rationality is bounded (Simon, 1955) by imperfect information, limited knowledge, and time constraints which can lead to decisions that are less than rationally optimal. A study shows that the decisions we make are also impacted by impulses, as "although the cold logic of self-interest is seductive, our first impulse is to cooperate" (Rand et al., 2012) for the convenience of maintaining peaceful interactions in everyday life. The fact that we are rational beings does not guarantee that we will always make moral choices, as our choices are also influenced by external circumstances as well as by the workings of our conscience, defined as "the person's interior judge of right and wrong" (Vischer, 2010), or as Adam Smith called it "the impartial spectator" or "the man within the breast" or the "tribunal of [one's] own conscience" (Smith, 1761). While neuroanatomical research has provided quite extensive knowledge of human consciousness, defined as "the function of the human mind that receives and processes information, crystallizes it and then stores it or rejects it with the help of the following: 1. The five senses 2. The reasoning ability of the mind 3. Imagination and emotion 4. Memory" (Vithoulkas and Muresanu, 2014), we have very little knowledge about the nature, origin, and the way human conscience acts. In 1761, Adam Smith, who was a deist, suggested that the "impartial spectator's" absolute standard of morality originated in the "all-seeing Judge of the world" (Smith, 1761); however, Smith's insight about conscience are mostly ignored, as they do not conform to science's universal view that there are no transcendental truths and that humans and their passions are just a meaningless accident. A more recent study presumes that conscience acts "not as a force that binds our will in particular circumstances, but more broadly as a set of truth claims that is perpetually in dialogue with our will" (Vischer, 2010). This view assumes that humans have free will, an ability which must be considered epistemically basic (Van Inwagen 1983) if humans' moral responsibility is presumed, as humans cannot be held morally responsible if their behaviour is predetermined and they are not free to choose. Through this line of reasoning, all humans inherently have the "moral law within", a fact revealed by the awareness of their existence (Kant, 1788/1998) and by the normal reactions of protest when someone is wronged in any way. Studies show that a moral sense is also prewired in preverbal infants and small children (Hamlin et al. 2007; Vaish et al., 2010), and this moral sense must be cultivated to flourish into morally conscious behaviour. While this may be a happy outcome, there is no guarantee that any human being will just make moral choices during their lifetime. This is due to the fact that, indeed, human beings are free to choose their courses of action (Cohen, 1995; List, 2019), and make either right or wrong choices in their actions, but they are not free to bear or not the consequences of those choices. Those consequences must be accepted, and their impacts are not limited to the person making the choice but they are often spread within society in space and time. Aside from having reason and moral sentiments, humans are also naturally "self-centred, self-interested and power-hungry creatures, concerned above all with their own preservation" as described by Thomas Hobbes in the seventeenth century (Hobbes, 1651/1994). Today, psychology research confirms that humans act in their social world according to the vertical moral continuum of the "great chain of being" (Lovejoy, 1936), as they "perceive others and themselves along the continuum from devilish to divine" (Brandt and Reyna, 2011) and having "emotions that attach us to saints and demons" (Haidt and Algoe, 2004). Indeed, by their nature, humans are complex beings, who can, at the same time, be selfish and generous, aloof and emphatic, hateful and loving, dishonest and honest, disloyal and loyal, cruel and kind, arrogant and humble, but most feel a little guilt over an excessive display of the first member of each of these seven pairs.

(Kagan, 2004)

Research indicates that indeed humans experience both guilt and shame, and that "feelings of guilt are apt to prompt apologies and reparation, whereas feelings of shame are apt to prompt denial and escape" (Tangney et al., 2007).

The above picture of human beings' complex and contradictory behaviours shows that human beings, even if endowed with reason, moral sentiments, and free will, are far away from achieving their telos of "wholeness" which Aristotle considered as the condition for human beings to participate in the maintenance of the wholeness of the life-sustaining Universe.

Why humans are not sustainable

Let us investigate the widespread confusion among humans about how the world functions and what is the real meaning of sustainability. I consider that this confusion is the result of at least two causes: (1) the way "sustainability" is seen today mostly as the anthropocentric concept of "sustainable development"; and (2) due to our pervasive lack of coherence between the "outer world and what we say about it" (Ekardt, 2020). I will not explain again the difference between the "sustainability" and the "sustainable development" concepts, as I explained the difference in the first chapter. I will focus instead on the second cause. At its core, the lack of coherence is due to a fragmented view of reality that disciplinary science has imposed on both our modern and postmodern thinking and which is now paving the way to an extremely reductionist view of reality, a view which sees life as technical data points rather than wholeness. According to Yuval Harari, "Science is converging to an all-encompassing dogma, which says that organisms are algorithms, and life is data processing" (Harari, 2015). Can humans still hope to survive in this world of super artificial intelligence and to achieve their sustainability aspirations?

Modernity and the dangers of scientism

In general, people respond to life's challenges according to their worldviews, defined as "the constellations of beliefs, values, and concepts that give shape and meaning to the world a person experiences and acts within" (Norton, 1991). I would argue that a particular worldview, namely that of *modernity*, has guided humans' prevalent ontological, epistemological, and axiological responses over the past three centuries, becoming the "universally" accepted paradigm of how the world functions, who humans are, what they can know, and what values they can hold. Started in Europe as a philosophical discourse after the revolutions of the seventeenth and eighteenth centuries and having since spread throughout the world, *modernity* is a concept of elusive definition. One definition refers to it as "a complex of more or less effectively realized concerns with power", where the concerns with power refer to "the capacity to change the structure of systems" (Elvin, 1986). According to Elvin, the systems whose structures have been mostly changed by modernity are two: (1) human beings – whether states, groups, or individuals – as well as (2) nature (Elvin, 1986). The general characteristics of modernity include: an emphasis on rationality and science over tradition and myth; a belief in progress and improvement; confidence in human mastery over nature; a focus on humanism, individuality, and self-consciousness; a close association to the birth and development of market capitalism; and a strong reliance on the state and its legal and governmental institutions (Linehan, 2009). Modernity has brought significant positive change in humans' lives during the Enlightenment and the Industrial Revolution, especially due to many scientific and technological advances, but also due to promotion of some novel social ideals of freedom, law, and justice, and the rise of representative democracy. However, modernity has also confined human thinking to a set of assumptions that have systematically undermined humans' ability to understand sustainability as the fundamental state of the Universe, and one which acts to support living beings in perpetuity. Packaged within modernity are assumptions that modern science, based on rationality and logical thinking, but devoid of values, can provide a correct picture of the complex, dynamic, interlinked social and ecological reality, as well as the right "formula" for humans' relationship with nature. In fact, modern science has given us a "natural ontology without a realist metaphysics" (Rolston, 2007), in which reality is "disenchanted" by "being denied meaning, significance and value" (Bhaskar, 2012).

Another modernity assumption is that the relative morality promoted by modernism, and deepened by the postmodern discourse developed after 1968 as a reaction to modernism (Bhaskar, 2012), can help humans get to a consensus on what sustainability is, and thus enable social action towards sustainability. By relative morality, I understand the feature of contemporary evaluative moral discourse and practice that reduces morality to personal preferences, feelings, or interpretations, by rejecting the existence of any objective and impersonal criteria, standards, or principles of moral judgement and action. By so doing, the "emotivist" moral judgements are "neither true nor false" and agreement "is not to be secured by any rational method [...]. It is to be secured, if at all, by producing certain non-rational effects on the emotions or attitudes of those who disagree with one" (MacIntyre, 1981).

To clarify why these assumptions are detrimental to sustainability, we need to explore modernity's evolution since the beginning of the Enlightenment in the seventeenth century. It was the French scholar Rene Descartes (1596–1650) who initiated the rationalist movement in science by famously declaring "*cogito ergo sum*" (I think, therefore I am). Obviously, Descartes was wrong believing that we can think before we exist and have the thinking mechanism (brain, mind) in place. However, his logical error of confusing cause and effect survived over centuries within Western thought with the dire consequence of excluding Aristotle's final cause (*telos*) from science and philosophy (Gilson, 2009). Thus, modern science paints an ontologically deficient view of the world in which nature is no longer seen as an ordered and *telos*-based whole, designed to provide a life-sustaining system for all living creatures, but as a fractured entity, where one species thinks that it can dominate the system and manage it as a "resource for technical exploitation and the construction of a human-created world" (de Vries, 2019). The modernity ethos has provided an excuse for humans to behave as if they own nature, by leading them to think

> that it is right for them to dominate the natural order and radically transform it into consumer goods, that it is necessary and acceptable to ravage the landscape in the pursuit of maximum economic production, and that only things produced by industry and placed on the market for sale have value.
>
> (Du Pisani, 2006)

The modern belief in the power of human reason to think abstract thoughts has led to numerous discoveries in natural sciences but, at the same time, it has sanctified a reductionist scientific method. For instance, science's "abstract universality, which finds its most striking manifestation in the forms of symbolic logic and money" has distorted the view of values by making money the dominant value in our world (Bhaskar, 2012). This has allowed the commodification of nature's values in order to use more efficiently biodiversity and ecosystem services. In a 2021 British report (Dasgupta, 2021), not only nature but the whole world is converted to financial assets, future generations are discounted, and nature that does not pay enough is liquidated as a bad investment (Spash and Hache, 2022).

Modern science starts with the fundamental presupposition that the world can be described by mechanistic reproducible models which are basically invariant as to time or place (Hutchinson, 2007). Based on empirical observation, the scientific method favours studying material facts and excludes other entities belonging to the metaphysical, studying, for instance, the brain but not the mind or conscience. Materialistic reductionism has hurt mostly modern biology which by adopting the mechanistic thinking of physics and chemistry has become just "an engineering discipline" unable to answer the fundamental questions about living organisms. Instead of explaining the evolution, emergence and innate complexity of the world of the living, biology aims to change the living organisms without trying to understand them (Woese, 2004). By ignoring substantive biological concepts concerning human beings' constitution and mode of functioning, such as "the concept of mind, of the human soul or of life" (Heisenberg, 1958), the reductionist materialism in biology has painted a dismal and unrealistic picture of human beings. This distorted picture of humans was widely spread by Charles Darwin's theory of random evolution which placed humans in the kingdom of animals. In 1871,

Darwin wrote in his book *The Descent of Man*: "The mental faculties of man and the lower animals do not differ in kind, although immensely in degree. A difference in degree, however great, does not justify us in placing man in a distinct kingdom" (Darwin, 1871). By blurring the distinction between humans and animals, Neo-Darwinists have maintained the myth that there is nothing unique in human beings. As I have tried to show, there is a human nature, with features necessarily specific only to humans, not to other biological species, or to humanoid robots (Marks, 2022). As David Berlinski wrote in 2019, "We are not simply apes with larger brains or smaller hands, and the distance between ourselves and our nearest ancestors is what it has always appeared to be, and that is practically infinite" (Berlinski, 2019). Darwin's erroneous thinking about humans' evolution from animals is still mainstream knowledge in science. His theory undermines humans' dignity and makes it impossible for humans to acquire real knowledge about themselves and about their place and responsibility within the systems of the world. Haakonssen has dubbed this real type of knowledge "system knowledge", a type of knowledge which deals with

> the understanding of things, events or persons in some sort of functional relationship to a greater "whole" or system – or understanding of individual actions in terms of their tendency, or "utility" to a view of all the elements of the universe as forming a teleological and perhaps divinely directed order.
>
> (Haakonssen, 1981)

Unbiased system knowledge is essential for understanding sustainability and how human beings can make correct choices motivated not only by self-interest but also by moral values, including ultimate moral values. However, modern science does not accept that ultimate moral values exist objectively, like primary colours, and that they can motivate human behaviour. This is the most consequential result of methodological reductionism in social sciences, established the moment when transcendental truth disappeared, and everyone became entitled to his/her own truth. It was the Scottish empiricist philosopher David Hume (1711–1776) who, in the eighteenth century, developed an ethical philosophy based on human nature, custom, and habit. According to Hume, humans do not gain awareness of good and evil rationally but through moral sentiments or feelings. His ethics does not require humans to be kind, compassionate or virtuous persons, they can display moral behaviour by just acting "humane". Hume defined "humanity as a feeling for the happiness of mankind, and resentment of their misery" (Hume, 1751/1998). He considered that this "humane" moral behaviour was required by the social condition of human life, and by humans' need to live in society, and, as a consequence of humans' dependence on each other to survive and satisfy their needs. This gave rise, in his opinion, to a "general sense of common interest. Which sense all the members of the society express to one another, and which induces them to regulate their conduct by certain rules" (Hume, 1888/1978). Hume explicitly names "utility" as the "true interest of mankind" and as the main source of moral approval. The tension between an individual's self-interest and the common interest can be solved, in Hume's opinion, by enforcing certain rules of justice, as means to secure approval or censure in human actions, leading the society towards progress, happiness, and the common good. A system of justice, for Hume, then, is ultimately nothing more than a "relatively settled set of conventional expectations between individuals chiefly concerned with their own interest that is reinforced by sentiments of approval and disapproval" (Hume, 1888/1978). Hume's naturalistic account of moral judgement has replaced the moral values based on objectively defined criteria of human behaviour with relativistic values based on human feelings and preferences, with the factual preferences of people being considered "right" *per se* (Ekardt, 2020). By rejecting the traditional teleological order that gives meaning to a moral human life, Hume has generalized a shallow conception of the human being which, being freed from the traditional morality, has adopted an "atomistic, individualistic, rights-and contract-centered moral thought" (Haakonssen, 1995), which made humans feel sovereign in their moral authority. Thus, Hume's moral theory has "under-cut the power of aspiration as a source of agency" (Krause, 2004) and has largely undermined the individual's wholeness understood as "connectivity to ourselves, to each other, and to the environment" (Fiscus and Fath, 2019). This has left the autonomous individual prone to engage in manipulative practices in order to protect his/her autonomy (MacIntyre, 1981) and promote empty "rights talk" (Schmidt, 2000) unconnected to responsibilities. "If rights are all that is important, what will become of responsibilities? If the individual man is the measure of all things, what will become of the family that produces and defines man?" (Wilson, 1993). Moreover, the secularization of morality has blurred the true/false meaning of moral judgements, a trend which has flourished in the postmodern culture of emotivism and judgemental irrationalism, manifested as "a denial of the possibility of giving better or worse grounds for a belief" (Bhaskar, 2012). Left without an ultimate moral authority to guide behaviour options, and without a teleological framework for ethics, how can the modern individual person be expected to make right decisions, including sustainable choices?

Relative morality

Two ethical frameworks have been proposed to replace the traditional morality, namely utilitarianism and deontology, which have had a huge social impact by transforming both theories and actions and reconstructing the social order. Utilitarianism is based on Jeremy Bentham's (1748–1832) empirical philosophy that the new human *telos* is to choose between those human actions that maximize pleasure (utility) and minimize pain. Deontology is based on Immanuel Kant's (1724–1804) philosophy that rational humans have the duty of right action. For Kant, the rules of morality leading to right action have their origin neither in human feelings nor in God's moral law, but in rational humans' practical reason, as "it is of the essence of reason that it lays down principles which are universal, categorical and internally consistent" (MacIntyre, 1981) for all humans. Finding "a maxim which all rational beings might will to accept as a universal moral law" was the challenge that Kant had faced (Kant, 1785/1998, 1788/1998). Kant found such a maxim to be the "principle of respect for persons" (Hodson, 1983), namely "always act so as to treat humanity whether in your own person or in that of others, as an end, and not as a means" (Kant, 1785/1998). However, Kant's ethical theory was doomed to fail as a "categorical imperative", as humans' rationality is not the best guide to moral action; besides, humans are not very fond of rule following. In fact, Kant himself recognized the weakness of his ethical theory when "in the second book of the second Critique, he does acknowledge that without a teleological framework the whole project of morality becomes unintelligible" (MacIntyre, 1981).

In the field of economics, utilitarianism and deontology have produced a value-free theory of choice which sees individuals as perfectly rational beings whose goal is to maximize their own utility (wellbeing) defined as "pleasure, desire fulfilment and choice" (Sen, 1982). Individuals do not make interpersonal utility comparisons and are "indifferent" in their choices. However, through the workings of a perfectly competitive market system, allocative efficiency (welfare) seems to occur as by magic. According to Samuelson and Nordhaus, "when each producer maximizes profits and each consumer maximizes utility, the economy as a whole is Pareto efficient" (Samuelson and Nordhaus, 1992), meaning that you cannot make anyone better off without making someone else worse off. As actual Pareto efficiencies are exceptionally rare, economists resorted instead to seeking potential Pareto improvements in the Kaldor–Hicks sense: welfare results if the magnitude of gains and the magnitude of losses are such that the gainers can fully compensate the losers for their losses and still be better off themselves (Kaldor, 1939; Hicks, 1940). By applying this philosophy to the income redistribution process, the government's role has increased in classical liberalism to a "welfare state" able to secure every individual equal economic opportunities and impartiality by applying John Rawls' minimalist principle of justice. This principle prescribes that every public decision should benefit the least-advantaged in society (Rawls, 1971). It is an irony that Vilfredo Pareto, who tried hard to get rid of external benchmarks for assessing social moral goodness in order to establish economics as a positive science, is considered "the father" of "Welfare Economics" defined as "the study of what is right and what is wrong, what is desirable and what is undesirable about the economy functioning" (Pareto, 1909/1976). Another Pareto "innovation" has been to replace the classical term "individual/group utility" with the Greek term "ophelimity" (useful or serviceable) "to designate the relationship of convenience which makes a thing satisfy a need or a desire, *whether legitimate or not*" (Pareto, 1909/1976). By introducing the concept of human desire or want, meaning (unlimited) satisfaction, as compared to human need (objectively limited) satisfaction, the theory of choice in economics justifies the goal of unlimited economic growth as the only developmental path to progress and social wellbeing. This unilinear path of development was generalized in 1959 by W.W. Rostow in his theory of the "stages of economic growth", as stages that all countries must follow in order to modernize (Rostow, 1959). Details about sources of economic growth were given in the theoretical model of long-run economic growth developed in 1956 by economists Robert Solow (1956) and Trevor Swan (1956). The model, which is still used by most countries seeking to develop, has as variables just two factors of production – man-made capital and labour – and a residual contribution of total factor productivity (TFP), but it assumes away natural resources, such as scarce raw materials. By assuming insatiable human wants and infinite natural resources able to allow unlimited economic growth, the model paints an unrealistic picture of humans as hybrid creatures who can live life autonomously from the chain of nature. A human is seen "as a spirit creature existing independently of anything material that lives on an endlessly usable earth" (Ekardt, 2020). Under the Western modernity worldview, humans' freedom has been limited to an "economically understood freedom" (Ekardt, 2020) to maximize income and profits, as if human beings could enjoy life without clean water and air, healthy food, and without feeling useful to others and being welcome as "persons in community" (Daly and Farley, 2011), with dreams and expectations to "stay alive and keep a place under the social sun" (Georgescu-Roegen, 1971). After the Second World War, increased economic growth has led, mainly in the advanced industrial economies, to the consolidation of modern technological societies. The main institutional pillars of these societies are an elaborate bureaucracy, a rational legal system enforcing the law of contracts, a scientific elite promoting "science as a source of awe", and relatively free markets (Berlinski, 2019). The increased economic security in these societies has produced a significant cultural shift, namely a greater emphasis on the "Post-Materialistic" values of individualism, autonomy, and self-expression, as well as a systematic erosion of religious practices, beliefs, and values (Inglehart, 2018). Globalization and technological change, which have accelerated in recent decades, have made great strides in diminishing global poverty. According to World Bank data, in 2020 about 700 million people, representing 9.3 per cent of the world population, were living in extreme poverty – defined as those who live on less than $2.15 per person per day at 2017 purchasing power parity (World Bank, 2022). However, both globalization and technological change have led to growing income and wealth disparity. "In 1965, the CEOs of major corporations in the USA were paid 20 times as much as their average employee. By 2012, they were paid 354 times as much" (Inglehart, 2018). In 2012, Joseph Stiglitz calculated that the upper 1 per cent of Americans controlled 40 per cent of the nation's wealth, and predicted no social peace, as in the future the central economic conflict will be "between the 1% and the 99%" (Stiglitz, 2012). In addition, the artificial intelligence (AI) revolution has made possible a new type of development leading to an Artificial Intelligence society which "will come into full force within the next twenty years as did the digital one since 1995 and will probably have an even greater impact than both the Industrial and digital ones combined" (Makridakis, 2017). Some believe that in this future society, life as we know it will disappear, as computer scientists having acquired the knowledge of how to inscribe intelligence on inorganic matter will create new forms of life based on conscious-free biochemical algorithms, with potential to create unprecedented social inequality by concentrating the new data-based wealth (Harari, 2015).

Humans' need for wholeness

The Western modernity worldview has been epistemologically challenged by new discoveries in science (Nagel, 2012; Nielsen et al., 2020), and currently is not shared globally, "as many people still live in a premodern world in which traditional morality and ethics prevail" (de Vries, 2019). In 1990, V.S. Naipaul, a Hindu by birth and a great English-language novelist, had described in an essay entitled *Our Universal Civilization* the turmoil of people like him torn between giving up their native culture based on "ritual and sacred texts" in order to adopt the Western values of what he perceived as the universal civilization. He mentioned the sense of "emptiness" and "alienation" that Muslim and Hindu people could feel in the Western world, as while their sacred texts "were cultural markers, giving us a sense of the wholeness of our world and the alienness of what lay outside", the universal civilization required them to adopt "ambition, endeavour, individuality" as main values (Naipaul, 1991). This represents a serious challenge for implementation of the 17 UN global Sustainable Development Goals (UNDP, 2015), which were developed using late-modernity assumptions. The 17 aspirational goals aim to achieve by 2030 social, economic, and environmental sustainable development worldwide, by integrating "people, planet and prosperity". However, the SDGs are based on the faulty assumption that economic growth and technological innovation, the very tools humans have used with impunity for beating nature and sustainability into submission, are the engines of sustainable development. A new study has assessed that the SDGs "fail to recognize that planetary, people and prosperity concerns are all part of one earth system, and that the protection of planetary integrity should not be a means to an end, but an end in itself" (Biermann et al., 2022).

What can be done?

Recognizing the need to protect planetary integrity is one step in the direction of correctly understanding sustainability, but I believe that changing humanity's unsustainable course will require a lucid introspection of ourselves and targeted efforts to rebuild the wholeness that we have lost when, through hubristic choices, we have stepped outside of the great circle of being that is our sustaining cosmos. Emboldened by an extreme egotistic view of ourselves, proud of our mentally constructed modern identities, which justify claiming human rights over human responsibilities, we have built an artificial world in which we do not feel free, being estranged from ourselves, from each other, from the natural world, and from "the higher-dimensional ground" which is able to heal our fragmented world (Bohm, 1980). Then many people are unable to see meaning in life and give in to the desire for immediate self-gratification by shopping, preferring man-made tanning beds to vitamin D-providing sunrays, favouring money over life when they choose to work three jobs a day, choosing sex over love, and choosing to offer medically assisted dying to suffering people instead of giving them loving care. In 1954, Einstein perceptively diagnosed that humans are the cause of our predicament, as with our inflated egos we are endangering our survival as a species. He suggested that we need "a substantial new manner of thinking" about ourselves if we want humanity to survive:

> A human being is part of a whole, called by us the universe, a part limited in time and space. He experiences himself, his thoughts and feelings, as something separate from the rest, a kind of optical delusion of his consciousness. This delusion is a kind of prison for us, restricting us to our personal desires and to affection for a few persons nearest us. Our task must be to free ourselves from this prison by widening our circles of compassion to embrace all living creatures and the whole of nature in its beauty. The true value of a human being is determined by the measure and the sense in which they have obtained liberation from the self. We shall require a substantially new manner of thinking if humanity is to survive.
>
> (Einstein, 1954)

Critical realism asserts the "inexorability of ontology" (Bhaskar, 2012) in understanding human beings and their actions. This means that humans should be analyzed not only as modernist abstractions, but as complex material and spiritual beings. According to critical realism, any human being exists stratified in three ontological levels: the ego (the illusionary sense of a self that is separate from other beings and from the being itself), the actual self (the embodied personality, as it exists and acts in real life), and the transcendentally real self. The real self, also called a "ground-state", is a human's innermost being, which endures and acts as a potentiality caused by the human's belonging to our interconnected Universe, and which "is the source of our creativity, our love, our genius, that is the source of everything which is good and noble that human beings do" (Bhaskar, 2012). In the real self, we discover a "more comprehensive, deeper, and more inward actuality" which is "neither mind nor body but rather a yet higher-dimensional actuality, which is their common ground and which is of a nature beyond both"

(Bohm, 1980). In our competitive Western society, the ego, that part of the self that Einstein tasked us to free ourselves from, has evolved from self-confidence to self-esteem and has flourished into egocentric or narcissistic attitudes, while the actual self has blossomed into numerous, fluid, multifaceted network identities (Wallace, 2019), which might be social, professional, gender, or nationality-based. In the meantime, the real, alethic, true self is ignored, or confused with the actual self, though sometimes, studies based on neuro-psychological research acknowledge its existence by mentioning decisions made by the self in the "conscious and unconscious brain" (Tversky and Kahneman, 1974). I propose that the real self has the potential to free us from the prison of our unsustainability. How? By letting our conscious mind, Adam Smith's "impartial spectator", and our unconscious mind, where ultimate moral values are stored, guide us, and control our impulses, passions, and sentiments. This conscious choice will change us from within, up to a point when the Golden Rule "do unto others as you would have them do unto you" will make sense and will free us to become whole beings, connected to the universal wholeness of life. This change has the potential to impact all social actions of an individual in the four dimensions of "our material transactions with nature, our social interactions with others, the level of social structure and the level of the stratification of our own embodied personalities" (Bhaskar, 2012).

The change at the level of the individual can lead to change in society, according to the Transformational Model of Social Action (Bhaskar, 1979). This model (Figure 4.1) assumes that all humans are social beings characterized by irreducible agency and intentionality, as well as by spontaneity of action. All human beings are born in a society which pre-exists them and provides the conditions (language, institutions, practices, values, infrastructure, conventions) necessary for their intentional activity. We can say that society does not exist independently of human activity, but it is not the product of this activity (Bhaskar, 1979).

The process whereby a society acts on the individual while he/she acquires the skills, practices, and habits necessary for living in that specific society and for being able later to reproduce or transform it is called "socialization". Various institutions, such as the family, the school, the sports venue, the church, are involved in the socialization process by providing both the theoretical and practical knowledge and the environment for healthy development

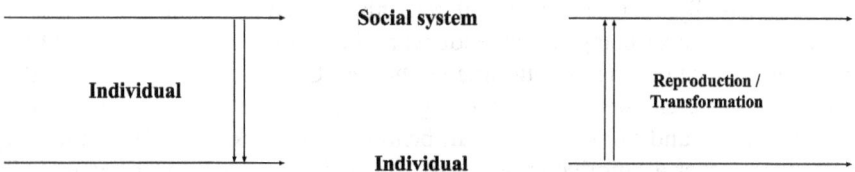

Figure 4.1 The Transformational Model of Social Activity
Source: Bhaskar, 1979.

of the socialized actual and real self. As for the social activity of reproduction/transformation, we should see it "as one in which people self-consciously transform their social conditions of existence (the social structure) so as to maximize the possibilities for the development and spontaneous exercise of their natural (species) powers" (Bhaskar, 1979). A critical mass of individuals who deliberately choose to focus not on their egos but on the potential for change existing in their true real selves will turn a society from unsustainable ways to sustainable ways of existing, thinking, and acting. The model thus can explain change toward sustainability because the change in individual agents can produce intentional change in social structures, which are by their nature transitory, given that society is an open system.

We have seen that the narrative of the UN SDGs based on late-modernity principles is not shared globally. Numerous religious people in both Western and Eastern traditions, as well as Indigenous peoples, do not recognize themselves in the dominant narrative of the SDGs which recommends that "the world should and will, in the name of universal human rights and the desirability of a modern lifestyle, follow a path of continuous material prosperity (measured as GDP per capita)" (de Vries, 2019), and consequently should transition from community-based and traditional rules and practices towards a modern welfare state. These people have different perspectives on what constitutes the good life, and have organized their community life based on traditional values, such as justice, temperance, and self-control, and on sustainable practices which promote living in harmony with nature and with each other. They have rich ontologies which include both seen and unseen elements. For instance, "in the African view, the universe is both visible and invisible, unending and without limits [...] man is not the master of the universe [...] he must live in harmony with the universe, obeying the laws of natural, moral and mystical order" (Mebratu, 1998). In the Chinese classical thought rooted in the Chinese agricultural civilization and in Confucianism, there is harmony and unity between heaven, Earth, and humans as "humans coordinate with heaven, with Earth, and heaven with Earth" (Liu, 2020). In this realistic view of life on Earth, human experiences include unexpected natural disasters such as earthquakes, tsunamis, hurricanes, volcanic eruptions, and other "acts of God", as well as miraculous healings and survivals from accidents. While these events may find natural explanations, it is impossible for humans to always understand based on science only why they happen or to predict their occurrence. For instance, science cannot explain why some organs in the human body provide useful functions but also pleasure. Such is the skin which primarily protects the body but is also sensitive, allowing humans to enjoy the touch of sunshine and of rain, or the digestive system which is fit to nourish bodies while at the same time providing the enjoyment of taste. Science has not found a rational explanation for the appearance of the first living cell on Earth, and cannot predict the time of death of any creature. This failure of science to explain and to predict future events, as well as to confirm hypotheses concerning human life leads us to infer that ontic openness also exists in the social realm. This possibility led Karl Popper to introduce the concept of "propensity interpretation of probability" in his falsification theory of science: "instead of speaking of the possibility of an event occurring, we might speak, more precisely, of an inherent propensity to produce, upon repetition, a certain statistical average" (Popper, 1990). Taking into account ontic openness, accepting the possibility that external forces more powerful than humans impact life on Earth, is not a sign of weakness but of strength, which could lead humans to innovation (Lane and Maxfield, 2004) and more sustainable choices.

The outbreak of the COVID-19 virus that produced the 2020 global pandemic is a case in point. Even if some researchers at Hong Kong University had predicted, based on data concerning a previous coronavirus epidemic (SARS), that a pandemic from a novel coronavirus would be likely to occur in the future, due to the "presence of a large reservoir of SARS-CoV-like viruses in horseshoe bats, together with the culture of eating exotic mammals in southern China" (Cheng et al., 2007), science could not accurately predict when the pandemic would be over, how many people would be infected, and what the best policies to combat the virus were. Like sustainability, the virus exists objectively, but the way people respond to this objective occurrence depends on what people consider most meaningful. The fact that most countries decided to protect human life from the harmful virus and locked down economic systems produced temporary favourable outcomes for environmental sustainability and demonstrated that other development paths are possible not only based on continuous economic growth. A study showed that in a few short weeks in the summer of 2020, the global response to the COVID-19 virus reduced greenhouse gas (GHG) emissions in China by an estimated 25 per cent, caused a 50 per cent reduction in nitrogen oxides in California, and visibly reduced NO_2 levels over Italy and China (Worstell, 2020).

In conclusion, sustainability exists objectively in all human societies as an alethic truth; however, it is not recognized as such due to humans' limited understanding of reality and to the socio-cultural-political impact of the modernity narrative, which has given rise to a dogmatic universal theory of how the world functions and how the autonomous modern self relates to it. Globalization and the IT revolution contribute to the spreading of the modernity narrative with potential to deepen social inequalities and to further destroy the environment. The critical realist Transformational Model of Social Activity is proposed as a potential way to bring change towards sustainability at the individual level. This change starts with the recognition that real human selves with tremendous intellectual and moral potential are part of the global sustainability project, a project which can be implemented if we focus on some significant ethical dimensions which are currently missing from the "sustainable development" project. To discuss these is the aim of the next chapter.

References

Adeney, T.J. (2022) *Altered Earth Getting the Anthropocene Right*. Cambridge: Cambridge University Press.

Ayala, F.J. (2010) "The difference of being human: Morality." *PNAS*, 107 (supplement_2): 9015–9022. https://doi.org/10.1073/pnas.0914616107.

Barrow, J.D. and Tipler, F.J. (1986) *The Anthropic Cosmological Principle*. Oxford: Clarendon Press; New York: Oxford University Press.

Berlinski, D. (2019) *Human Nature*. Seattle: Discovery Institute Press.

Bhaskar, R. (1975) *A Realist Theory of Science*. Leeds: Leeds Books.

Bhaskar R. (1979) *The Possibility of Naturalism: A Philosophical Critique of the Contemporary Human Sciences*. Atlantic Highlands, NJ: Humanities Press.

Bhaskar, R. (2012) *Reflections on MetaReality: Transcendence, Emancipation and everyday Life*. London and New York: Routledge.

Biermann, F., Hickmann, T., and Sénit, C. (eds) (2022) *The Political Impact of the Sustainable Development Goals: Transforming Governance Through Global Goals?* Cambridge: Cambridge University Press.

Bohm, D. (1980) *Wholeness and the Implicate Order*. London and New York: Routledge.

Brandt, M.J. and Reyna, C. (2011) "The Chain of Being: A Hierarchy of Morality." *Perspectives on Psychological Science*, 6(5): 428–446.

Cheng, V.C.C., Lau, S.K.P., Woo, P.C.Y., and Yuen, K.Y. (2007) "Severe Acute Respiratory Syndrome Coronavirus as an Emerging and Reemerging Infection." *Clinical Microbiology Reviews*, 20(4), 660–694. https://doi.org/10.1128%2FCMR.00023-07.

Cohen, H. (1995) *Religion of Reason*. Atlanta, GA: Scholars Press.

Daly, H. and Farley, J. (2011) *Ecological Economics: Principles and Applications*. Washington, DC: Island Press.

Darwin, C. (1871) *The Descent of Man, and Selection in Relation to Sex*. London: John Murray.

Dasgupta, P. (2021) *The Economics of Biodiversity: The Dasgupta Review*. London: HM Treasury.

de Vries, B.J.M. (2019) "Engaging with the Sustainable Development Goals by going beyond Modernity: An ethical evaluation within a worldview framework." *Global Sustainability*, 2: 1–14.

Du Pisani, J.A. (2006) "Sustainable development – historical roots of the concept." *Environmental Sciences*, 3(2): 83–96.

Einstein, A. (1954) *Ideas and Opinions*. New York: Crown.

Ekardt, F. (2020) *Sustainability Transformation, Governance, Ethics, Law*. Cham, Switzerland: Springer.

Elvin, M. (1986) "A Working Definition of 'Modernity'." *Past and Present*, 113: 209–213.

Ferraris, M. and Torrengo, G. (2014) "Documentality: A Theory of Social Reality." *Rivista di estetica*, 57: 11–27. https://journals.openedition.org/estetica/629#ftn3.

Fiscus, D.A. and Fath, B.D. (2019) *Foundations for Sustainability. A Coherent Framework of Life-Environment Relations*. London: Elsevier Academic Press.

Georgescu-Roegen, N. (1971) *The Entropy Law and the Economic Process*, Cambridge, MA: Harvard University Press.

Georgescu-Roegen, N. (1979) "Comments on Papers by Daly and Stiglitz." In *Scarcity and Growth Reconsidered*, edited by K.V. Smith (2011). New York: NY: Resources for the Future Press.

Gilson, E. (2009) *From Aristotle to Darwin and Back Again: A Journey in Final Caus-ality, Species and Evolution*, translated by J. Lyon. San Francisco, CA: Ignatius Press.

Gough, I. (2017) *Heat, Greed and Human Need*. Cheltenham, UK/Northampton, MA: Edward Elgar.

Haakonssen, K. (1981) *The Science of a Legislator: The Natural Jurisprudence of David Hume and Adam Smith*. Cambridge: Cambridge University Press.

Haakonssen, K. (1995) *Natural Law and Moral Philosophy: From Grotius to the Scottish Enlightenment*. Cambridge: Cambridge University Press.

Haidt, J. and Algoe, S. (2004) "Moral amplification and the emotions that attach us to saints and demons." In *Handbook of Experimental Existential Psychology*, edited by J. Greenberg, S. L. Koole, and T. Pyszczynski. New York: Guilford, pp. 322–335.

Hamlin, J.K., Wynn, K., and Bloom, P. (2007) "Social evaluation by preverbal infants." *Nature*, 450: 557–559.

Harari, Y.N. (2015) *Homo Deus. A Brief History of Tomorrow*. Toronto: Signal Pen-guin Random House.

Heisenberg, W. (1958) *Physics and Philosophy: The Revolution in Modern Science*. New York: Prometheus Books.

Hicks, J. (1940) "The Valuation of Social Income." *Economica*, 7: 105–124.

Hobbes, T. (1651) "Leviathan." In *Leviathan, with selected variants from the Latin edition of 1668*, edited by E. Curley (1994). Indianapolis: Hackett.

Hodson, J.D. (1983) "The Ethics of Respect for Persons." In *The Ethics of Legal Coercion. Philosophical Studies Series in Philosophy*, vol. 26. Dordrecht: Springer. https://doi.org/10.1007/978-94-009-7257-5_1.

Hume, D. (1998) *An Inquiry Concerning the Principles of Morals*, edited by T.L. Beauchamp. Oxford: Oxford University Press. (Original work published 1751)

Hume, D. (1978) *A Treatise of Human Nature*, edited by L.A. Selby-Bigge (2nd edi-tion). Oxford: Clarendon Press. (Original work published 1888)

Hutchinson, I.H. (2007) "Warfare and Wedlock – Redeeming the Faith-Science Rela-tionship." *Perspectives on Science and Christian Faith*, 59(2): 91–101.

Inglehart, R.F. (2018) *Cultural Evolution People's Motivations are Changing and Reshaping the World*. Cambridge: Cambridge University Press.

Kagan, J. (2004) "The uniquely human in human nature." *Daedalus*, 133(4): 77–88.

Kaldor, N. (1939) "Speculation and Economic Stability." *The Review of Economic Studies*, 7(1): 1–27.

Kant, I. (1998) *Grundlegung zur Metaphysik der Sitten*. Frankfurt: Suhrkamp. (Origi-nal work published 1785)

Kant, I. (1998) *Kritik der praktischen Vernunft*. Frankfurt: Suhrkamp. (Original work published 1788)

Kant, I. (1997) *Critique of Practical Reason, Cambridge Texts in the History of Phi-losophy*. Cambridge: Cambridge University Press. (Original work published 1788)

Krause, S.R. (2004) "Hume and the (false) luster of justice." *Political Theory*, 32(5): 628–655.

Lane, D.A. and Maxfield, R.R. (2004) "Ontological Uncertainty and Innovation." *Journal of Evolutionary Economics*, 15(1): 3–50.

Leitenberg, M. (2006) "Deaths in Wars and Conflicts in the 20th Century" (3rd edi-tion). Cornell University Peace Study Program, Occasional Paper No. 29.

Linehan, D. (2009) "Modernity." *International Encyclopedia of Human Geography*. www.sciencedirect.com/topics/earth-and-planetary-sciences/modernity.

List, C. (2019) *Why Free Will Is Real*. Cambridge, MA: Harvard University Press.

Liu, Y. (2020) "The Paradigm of the Wild, Cultural Diversity, and Chinese Environmentalism: A Response to Holmes Rolston III." *Environmental Ethics*, 42(3): 223–235.

Lovejoy, A.O. (1936) *The Great Chain of Being*. Cambridge, MA/London: Harvard University Press.

Lynas, M. (2011) *The God Species: Saving the Planet in the Age of Humans*. Washington, DC: National Geographic.

MacIntyre, A.C. (1981) *After Virtue: A Study in Moral Theory*. Notre Dame, IN: University of Notre Dame Press.

Makridakis, S. (2017) "The Forthcoming Artificial Intelligence (AI) Revolution: Its Impact on Society and Firms." *Futures*, 90: 46–60.

Marks, R.J. (2022) *Non-Computable You What You Do That Artificial Intelligence Never Will*. Seattle, WA: Discovery Institute Press.

Mebratu, D. (1998) "Sustainability and sustainable development: historical and conceptual review." *Environmental Impact Assessment Review*, 18: 493–520.

Nagel, T. (2012) *Mind and Cosmos: Why the Materialist Neo-Darwinian Conception of Nature is Almost Certainly False*. New York: Oxford University Press.

Naipaul, V.V. (1991) "Our Universal Civilization." *City Journal*. New York: Manhattan Institute for Policy Research.

Nielsen, S.N., Fath, B.D., Bastianoni, S., Marques, J.C., Muller, F., Patten, B.C., Ulanowicz, R.E., Jorgensen, S.E., and Tiezzi, E. (2020) *A New Ecology Systems Perspective* (2nd edition). Amsterdam/Cambridge, MA: Elsevier.

Norton, B.G. (1991) *Toward Unity Among Environmentalists*, New York: Oxford University Press.

Pareto, V. (1976) *Sociological Writings*, selected and introduced by S.E. Finer. Oxford: Basil Blackwell. (Original work published 1909)

Polanyi, M. (1974) *Personal Knowledge. Towards a Post-Critical Philosophy*. Chicago: University of Chicago Press. (Original work published 1958)

Popper, K. (1990) *A World of Propensities*. Bristol: Thoemmes Antiquarian Books.

Princen, T. (2010) *Treading Softly: Paths to Ecological Order*. Cambridge, MA: MIT Press.

Rand, D.G., Greene, J.D., and Novak, M.A. (2012) "Spontaneous Giving and Calculated Greed." *Nature*, 489: 427–430.

Rawls, J. (1971) *A Theory of Justice*. Cambridge, MA: Harvard University Press.

Rolston, H. III (1989) *Philosophy Gone Wild*. Amherst, NY: Prometheus Books.

Rolston, H. III (2007) "Living on Earth: Dialogue and Dialectic with my Critics." In *Nature, Value, Duty*, edited by C.J. Preston and W. Ouderkirk. Dordrecht: Springer, pp. 237–268.

Rolston, H. III (2017) "Technology and/or Nature: Denatured/Renatured/Engineered/ Artifacted Life?" *Ethics & the Environment*, 22(1): 41–62.

Rostow, W.W. (1959) "The Stages of Economic Growth." *The Economic History Review*, 12(1): 1–16.

Sachs, J. (1995) *Aristotle's Physics: A Guided Study (Masterworks of Discovery)*. New Brunswick, NJ: Rutgers University Press.

Samuelson, P.A. and Nordhaus, W.D. (1992) *Economics* (14th edition). New York:, McGraw-Hill.

Schmidt, J. (2000) "What Enlightenment Project?" *Political Theory*, 28(6): 734–757.

Schrödinger, E. (1967) *What Is Life? & Mind and Matter*. Cambridge: Cambridge University Press.

Sen, A. (1982) "Rights and Agency." *Philosophy and Public Affairs*, 11: 7–13.

Simon, H.A. (1955) "A Behavioral Model of Rational Choice." *Quarterly Journal of Economics*, 69(1): 99–118.

Sinclair, D.A. with LaPlante, M.D. (2019) *Lifespan. Why We Age, and Why We Don't Have To.* New York: Atria Books.

Smith, A. (1761) *The Theory of Moral Sentiments* (2nd edition). London: A. Millar; Edinburgh: A. Kincaid and J. Bell.

Solow, R.M. (1956) "A Contribution to the Theory of Economic Growth." *The Quarterly Journal of Economics*, 70(1): 65–94.

Solzhenitsyn, A. (2020) *The Gulag Archipelago 1918–1956. An Experiment in Literary Investigation.* New York: Harper Perennial. (Original work published 1973)

Spash, C.L. and Hache, F. (2022) "The Dasgupta Review deconstructed: an exposé of biodiversity economics." *Globalizations*, 19(5): 653–676.

Stiglitz, J. (2012) *The Price of Inequality: How Today's Divided Society Endangers Our Future.* New York: W.W. Norton & Company.

Swan, T.W. (1956) "Economic Growth and Capital Accumulation." *Economic Record*, 32(2): 334–361.

Tangney, J.P., Stuewig, J., and Mashek, D.J. (2007) "Moral emotions and moral behavior." Annual Review of Psychology, 58: 345–372.

Tainter, J.A. (1988) *The Collapse of Complex Societies.* New York: Cambridge University Press.

Thorvaldsen, S. and Hössjer, O. (2020) "Using statistical methods to model the fine-tuning of molecular machines and systems." *Journal of Theoretical Biology*, 501: 110352.

Tversky, A. and Kahneman, D. (1974) "Judgment under Uncertainty: Heuristics and Biases." *Science*, 185(4157): 1124–1131.

UNDP (2015) "United Nations Development Program, Sustainable Development Goals." www.undp.org/sustainable-development-goals.

Vaish, A., Carpenter, M., and Tomasello, M. (2010) "Young children selectively avoid helping people with harmful intentions." *Child Development*, 81, 1661–1669.

Van Inwagen, P. (1983) *An Essay on Free Will.* Oxford: Oxford University Press.

Vischer, R.K. (2010) *Conscience and the Common Good. Reclaiming the Space Between Person and State.* New York: Cambridge University Press.

Vithoulkas, G., and Muresanu, D.F. (2014) "Conscience and consciousness: a definition." *Journal of Medicine and Life*, 7(1):104–108.

Wallace, K. (2019) *The Network Self. Relation, Process and Personal Identity.* New York: Routledge.

Washington, H. (2015) "Is 'sustainability' the same as 'sustainable development'?" In *Sustainability Key Issues*, edited by H. Kopnina and E. Shoreman-Ouimet. London and New York: Routledge.

Wilson, J.Q. (1993) *The Moral Sense.* New York: Free Press.

Woese, C.R. (2004) "A New Biology for a New Century." *Microbiology and Molecular Biology Review* 68(2): 173–186.

World Bank (2022) "Poverty." www.worldbank.org/en/topic/poverty/overview.

Worstell, J. (2020) "Ecological Resilience of Food Systems in Response to the COVID-19 Crisis." *Journal of Agriculture, Food Systems, and Community Development*, 9(3): 1–8.

5   Sustainability and what really matters

Sustainability, defined as a natural and mysterious organization of planet Earth's structure and functioning in a way that supports the continuation of life on Earth, has a profound ethical dimension, as it shows the intrinsic worth of the planet's living creatures that can enjoy freely the "gift of a working life-support system" (Murphy et al., 2021). We can analyze this ethical dimension under several aspects: the value of temporal and spatial continuity of life on Earth, the value of life itself, and the value of the unity/interaction/connectivity among various living things and their environment. The objectively existing make-up of the world requires a response from humans as the only earthly living creatures endowed not only with rational minds but also with moral sentiments and emotions. Humans' response or deliberation about the life-sustaining system and how human beings relate to it, being based on value judgements, adds another important ethical dimension to sustainability. The ethical dimension of sustainability has been recognized in the relevant academic literature since the 1980s, mostly in studies of environmental ethics (Taylor, 1981; Rolston, 1986, 1988, 1991; Callicott, 1989; Norton, 1996; Becker, 1997). The concept of sustainability has been used in the last four decades in international political statements such as the Brundtland Report (WCED, 1987), the Rio Declaration (UN, 1992), and the Johannesburg Declaration (UN, 2002), as a normative concept aiming to guide the long-term global actions of humans. However, with the exception of a few volumes discussing sustainability ethics (Becker, 2012; Kibert et al., 2012; Gale, 2018), the ethical dimension of sustainability tends to be downplayed by the current political discourse as well as by most mainstream research, both of which are mostly focused on practical approaches of how to make sustainability work under the three identified aspects of sustainable development – ecological, social, and economic sustainability. This chapter is a discussion about sustainability as an objective value, a gift that gives meaning to human existence on planet Earth, as no living creature has produced the life-sustaining system or is expected to pay for enjoying breathable air, clean water, and the blessings of heat and light provided by sunshine. We have seen in a previous chapter that the existence of our "Goldilocks"-type Universe with numerous contingent properties fine-tuned to support life on

DOI: 10.4324/9781003307587-6

Earth has puzzled physicists, cosmologists, and philosophers of science since the 1950s. There have been numerous attempts at explaining fine-tuning – for instance, by proposing the anthropic principle (Barrow and Tipler, 1986; Lewis and Barnes, 2016), or making design inferences (Dembski, 1998; Meyer, 2009). The focus of this chapter is on the way humans as humans, and not as scientists, relate to the existence of the life-support system. If sustainability is seen as a "gift of life", it should inform the way human beings make choices, frame norms and regulations, and organize economic and civic life. Understood as a mechanism that secures continuation of life on Earth, sustainability is an "existential problem, not [only] an environmental and social one" (Ehrenfeld, 2008). This fact forces us to reconsider who we as humans are, what our final ends might be and how we relate to the kind of thing the life-sustaining Earth is. Do we have only rights or also responsibilities as the only rational and moral species engaged in the life-support system? Can the individual self be sustainable, or are there some virtues that are needed in order for the individual to become a sustainable person? Along this line of thought, implementation of sustainability becomes less a question of scientific and technical fixes and more an ethical issue requiring a great moral reset, by replacing the established narrative about the modern autonomous individual self, free to live independently according to self-made moral rules and practices, with a novel narrative of a human being which is a person of character, ethically self-less, and actively and responsibly involved in the great "web of relationships by which a place and its creatures sustain a mutual life" (Berry, 2015).

To answer the above questions, this chapter engages in three philosophical debates which are ongoing, both within and without environmental ethics. The first two concern moral philosophy, and the third one the philosophy of science. In epistemology, we are arguing for moral objectivism against those who regard values as subjective. Moral objectivism is the idea that what is right or wrong, including truth, exists factually and does not depend on what anyone thinks it is right or wrong (Rescher, 2008) or true. According to moral objectivism, moral codes can be compared to each other against a set of universal facts and not against mores, or customs in a society. Objective morality acknowledges the *is/ought* dichotomy, the difference between positive statements (about what *is*) and normative statements (about what *ought to be*). The dichotomy was first identified by the Scottish philosopher David Hume in his *Treatise of Human Nature* published in 1739. Hume insisted that we cannot obtain the truth of a moral statement based on factual premises only (Hume, 1978/1888). For instance, we cannot say, "climate change is happening, therefore we ought to fight climate change", and expect that this moral "ought" derived from empirical considerations "can pass the tests of prescriptiveness and categoricalness" (Gewirth, 1973–1974) required for the sentence to produce effective outcomes. Moral subjectivism is the idea that morality (what is good and what is evil) is decided by the individual, and thus there is no end to a moral debate, debaters agreeing to disagree. Moral subjectivism holds that there are no objective, intrinsic values, and all values are subjective. According to philosopher Charles Taylor, this conception, dominant in Western civilization since the seventeenth century, has produced three *malaises*: a loss of meaning, due to increased individualism; a loss of *telos* ("the eclipse of ends"), due to the rise of instrumental reason; and a loss of freedom, due to dominance of the "individualism of self-fulfilment" (Taylor, 1991). The second position we advocate, against Jeremy Bentham's utilitarianism and Immanuel Kant's deontology, is virtue ethics, believing with Taylor and Anscombe that "the most complete conception of ethical and political life must be rooted in virtue ethics and positive liberty" (Barry, 2019). Virtue ethics is the moral philosophy stating that "the right action is the action that a virtuous person would characteristically perform in the circumstances" (Hursthouse, 1999) – for instance, by not sustaining consumerism or an excessive anthropocentric lifestyle. Virtue ethics is a growing field in normative environmental ethics (Kohak, 2000; Henning, 2006), promoting the idea that a virtuous person can embody a moral character and become a model for others. The key thing in virtue ethics is that it is *not* based on clearly defined moral principles; it is based rather on virtuous actions in incalculable and diverse situations (McGrath, 2023).

With regard to the philosophy of science, this chapter is based in critical realism, following Roy Bhaskar's and his followers' philosophy of realism. Critical realism sees reality as stratified and including a transcendental dimension that "operates independently of our awareness or knowledge of it" and which "does not wholly answer to empirical surveying or hermeneutical examination" (Archer et al., 2016). This way, critical realism is best positioned to analyze sustainability in both nature and in the social realm, as it represents

> an alternative paradigm both to scientistic forms of positivism concerned with regularities, regression-based variables models, and the quest for law-like forms; and also to the strong interpretivist or postmodern turn which denied explanation in favor of interpretation, with a focus on hermeneutics and description at the cost of causation.
>
> (Archer et al., 2016)

The choice of these moral and epistemic philosophical positions may produce objections concerning, among others, the scientific defence of the is/ought dichotomy, the often-repeated remark that moral ethics is culturally relative, as well as the virtue ethical assumption of an eternal, normative human nature, and social constructivism as science. However, understanding sustainability correctly, and the possible ways of promoting this normative goal in our post-virtue society, warrants the risk of being exposed to these objections which, at the end of the day, might lead to a fruitful dialogue.

This chapter will develop a virtue-based theory of a "sustainable person", attempting to answer the question "How should a person be in order to be sustainable?" A sustainable person must be a moral person. The theory assumes that the human being is a complex, rational, and relational being who is able to critically assess the various relationships they engage in by virtue of their earthly condition, namely "our material transactions with nature, our social interactions with others, the level of social structure and the level of the stratification of our own embodied personalities" (Bhaskar, 2012). Only once such a reconsideration takes place in human beings' minds and hearts can we hope that a new narrative – as well as new practices to support sustainability – will emerge and start to challenge and change the current deeply unsustainable "consumeristic" institutional and cultural fabric of modern societies which threatens the very existence of Western civilization. Society will be changed by a change in individual human beings one at a time.

In previous chapters, we have seen that sustainability, defined as the ability of planet Earth to subsist and provide a life-sustaining system for all living creatures, exists as a matter of fact both in nature and in human societies. Yet, why is sustainability an ultimate value, or something good in itself? At the core of sustainability are three structural and functional characteristics which make it inherently valuable: (1) the persistence, durability, or temporal and spatial continuity of the life-support system; (2) the goodness of life as a whole; (3) the unity of the biotic community, or the relational character of sustainability. I will discuss each of these characteristics in what follows.

Continuity

Sustainability appears as the mysterious condition which keeps us alive on planet Earth. It permanently invades our lives as a matter of conscious experience, as intimate as the act of breathing or the act of thinking. As such, sustainability is not a human creation. It is something complex that we discover in our existence, with biological aspects that are non-negotiable (we all breathe, eat, move, work, sleep, and die) and aspects that are non-mandatory and involve human choices, such as swimming in an ocean, enjoying the moonlight, hunting, growing crops, or having children. Our life is maintained due to the workings of numerous visible and invisible entities and laws acting in nature and in our bodies. They include laws, such as the law of entropy; processes, such as the process of growth and decay; and goods and services provided by nature's ecosystems, such as clean air, food, water purification, aesthetic beauty (Millennium Ecosystem Assessment, 2005), but also basic elements such as primary colours, matter, energy, time and space, as well as irreducibly complex biochemical systems, such as the blood-clotting system (Behe, 1996). Sustainability appears to be non-discriminatory, as the life-support system is freely available to all living creatures, not only to humans. However, only humans are able to make moral value judgments which means that unlike frogs, for instance, who do not have cognitive abilities enabling them to make moral choices, humans and their "human interactions with nature can never be morally neutral, but are under judgement: Does this interaction contribute to the flourishing of the whole?" (Cooper et al., 2016), or do our actions hurt the whole which sustains us? Time and space are the two coordinates that organize the chaos of earthly existence by setting limits to the life span of every creature and by not affording the gift of ubiquity to any creature. Ecologists consider that "by nature, time is evolutionary and irreversible, whereas the space is conservative and reversible" (Nielsen et al., 2020). Scientific research shows that, following the time's arrow, low forms of life, such as bacterial cells, have existed on Earth for more than 3.7 billion years in submarine-hydrothermal vents (Dodd et al., 2017) and that humans in the form of *Homo sapiens* have been around for about 286,000 years (Richter et al., 2017). It is assumed that life on Earth will continue as long as the sun, the main source of free energy (low-entropy radiation), will continue to shine. There is no consensus among scientists about how long this time frame is, though most agree that the total free energy available from sunlight is huge, approximately 1.2×10^{17}W per year, which is about ten thousand times larger than the 1.3×10^{13}W of yearly global power consumption from burning fossil fuels (Lineweaver and Egan, 2008). There is also agreement among most scientists that the entropy of the Universe increases and will eventually lead to the heat death of the Universe (Thomson, 1862) in about 5×10^9 years (Barrow and Tipler, 1986) when the sun's free energy is expected to be exhausted.

Life on Earth takes place in a thin layer called the ecosphere which occupies a region extending about 10 km from the sea level into the ocean depths and approximately the same distance up into the atmosphere (Nielsen et al., 2020). In this relatively small space, a huge biodiversity exists. Using biomass as a measure of abundance, and measuring biomass in gigatons of carbon existing in living creatures (with 1 Gt C = 10^{15} g of carbon), an ecological study (Bar-On et al., 2018) has assessed that the sum of the biomass across all taxa (units used in the science of biological classification) on Earth is ≈550 Gt C, of which ≈80 per cent (≈450 Gt C) is in plants, located mostly on land. The second major biomass component is bacteria (≈70 Gt C), constituting ≈15 per cent of the global biomass. Other groups, which account for the remaining percentage, are fungi, archaea, protists, animals, and viruses. In the marine environment, there are many microbes (≈70% of the total marine biomass) along with arthropods and fish (≈30%). Biodiversity loss due to humans' intensified economic activities is a constant concern, as measured by Vitousek et al. (1997), Wackernagel and Rees (1996), Rockström et al. (2009), and Steffen et al. (2015). In 2019, according to the Global Assessment Report of Biodiversity and Ecosystem Services issued by the Intergovernmental Science-Policy Platform on Biodiversity and Ecosystem Services (IPBES), the rate of acceleration of global biodiversity loss has increased, threatening with extinction around 1 million species (IPBES, 2019) which represents about 10 per cent of the total number of species estimated to exist on Earth. Aside from species loss, there is an increase in the number of domesticated animals and birds compared to their wild counterparts. In 2018, it was assessed that the combined biomass of livestock (≈0.1 Gt C) (mostly cattle and pigs), as well as of domesticated poultry (≈0.005 Gt C) (mostly chickens) was higher than the biomass of humans ($\approx$0.06 Gt C), and much higher than the combined biomass of wild mammals ($\approx$0.007 Gt C) and wild birds ($\approx$0.002 Gt C) (Bar-On et al., 2018). Assessing that "biodiversity is fundamental to human well-being and a healthy planet", as "it supports all systems of life on earth" (UN-CBD, 2022), and building on the 2011–2020 Strategic Plan for Biodiversity, the 15th Conference of the Parties to the Convention on Biological Diversity (COP 15), convened in Montreal in December 2022, adopted the Kunming-Montreal Global Biodiversity Framework, an ambitious plan with 23 action-oriented global targets to stop biodiversity loss by 2030, and prepare humanity "for living in harmony with nature" by 2050. Two of the targets expect all countries to accomplish, by 2030: (1) an effective restoration of at least 30 per cent of areas of degraded terrestrial, inland water, and coastal and marine ecosystems, and (2) the effective conservation and management of at least 30 per cent of terrestrial, inland water, and of coastal and marine areas, through "ecologically representative, well-connected and equitably governed systems of protected areas" (UN-CBD, 2022). These global actions for preserving biodiversity, echoing the 1987 moral commitment of the Brundtland Commission to secure conditions for sustainable development of future planetary generations, clearly demonstrate that continuation of life on Earth is valued by the global human community. "Our nature seems to require us to hope that our life and the world's life will continue into the future" (Berry, 2015). At the same time, no one knows the future, which is why using the precautionary principle in dealing with nature is a wise attitude to adopt, as well as continuing to learn more about the complex structure and functions of dynamic ecosystems, especially the complex interplay of "matter, energy and information" in biological systems (Nielsen et al., 2020).

Goodness of life

At the centre of the sustainability concept is life, not only as a theoretical concept but as lived by billions of creatures on Earth who experience it as a cycle (from being born, growing, living it, eventually reproducing it, and inevitably leaving it when the time comes). The cycle has a material component which closes when the body returns to dust (the earth), but also immaterial components, called reason, mind, awareness, consciousness, or soul, which characterize only human beings, as the only earthly creatures able to prefigure their own deaths. This is what makes human life *tragic*, but also magic, in a sense, forcing us to believe that there is meaning attached to life, beyond its material structure and function. What happens to the soul or mind at the physical body's death is not clear. Some say that nothing happens; our soul simply dies with the body (Martin and Augustine, 2015). Others figure that the soul continues to live eternally, being added to the cosmic mind (Masaeli and Sneller, 2018). Life as material structure and its functions have been thoroughly described by natural sciences, such as physics, chemistry, molecular biology, physiology, ecology, or systems sciences. For instance, life as living matter has been defined as "matter that shows certain attributes that include responsiveness, growth, metabolism, energy transformation, and reproduction" (Sagan, 2022). For some physicists, living beings are just material systems subject to the second law of thermodynamics: "[T]he one unequivocal thing we know about life is that it always dissipates energy and creates entropy in order to maintain its structure" (Anderson and Stein, 1987). For other scientists, living organisms are complex entities of a completely different kind than non-living inorganic matter: "Rather than an organism being just a standard material system plus a list of special conditions, an organism is a repository of meanings and impredicativities; it is more generic than an inorganic system rather than less" (Rosen, 2000). Most ecologists would agree that there is nothing simple about living systems. Being thermodynamically and ontically open, they "are highly heterogeneous and far more complex than physical systems. Their behaviour, as opposed to physical systems is nondeterministic and irreversible. This is what we today would designate as far from equilibrium systems or dissipative structures" (Nielsen et al., 2020). In other words, living beings are self-organizing systems able to purposefully keep themselves far from thermodynamic equilibrium (death) by feeding on negative entropy (while dissipating free energy) (Prigogine, 1980). Mere knowledge about these special structural and functional characteristics of living beings (plants, animals, and humans) should be enough to value them above the inanimate things in nature (rocks, iron, or coal). Do we? Not exactly. We call plants, animals, and marine species "renewable resources" because they renew themselves, but most of the time we overuse them by not giving them enough time to renew themselves. For instance, using the scientific measure of Maximum Sustainable Yield for harvesting fish, humans have tripled the percentage of wild fish stocks harvested at biologically unsustainable levels from 10 per cent in 1974 to 33.1 per cent in 2015 (Bailey and Tupy, 2020). New research into systems ecology crosses disciplinary boundaries and proposes that "life itself" (Rosen, 1991) should be considered the basis of sustainability value and ethics (Fiscus and Fath, 2019), for its potential to bridge the existing gap between humans and the biotic communities in the environment, leading to a "reverence for life" ethic (Schweitzer, 1965). Schweitzer believed that "reverence for life" means clearly distinguishing between "evil" which "is what annihilates, hampers, or hinders life" and "goodness, which, by the same token, is the saving or helping of life, the enabling of whatever life [one] can to attain its highest development" (Schweitzer, 1936) At the heart of Schweitzer's "absolute ethic of reverence for life" is the cosmological concept of the "will-to-live" or the instinct to preserve their own life which belongs to all living creatures. For Schweitzer, acting morally consists "in my experiencing the compulsion to show to all [other] will (s)-to-live the same reverences as I do to my own" (Schweitzer, 1987). This compassion and responsibility for all life receives mystical connotations for Schweitzer: "Whenever my life devotes itself in any way to life, my finite will-to-live experiences union with the infinite [W]ill in which all life is one" (Schweitzer, 1987).

Fiscus and Fath distinguish between "discrete life" (life lived by each individual living thing) and "sustained life" or "life as a unified whole inherent in communities, ecosystems and the biosphere". They acknowledge that our biosphere is organized to sustain life through the act of destruction of discrete life: "life itself is a continual process of killing, eating, dying, going extinct in individual organism/species form to sustain life as a unified whole" (Fiscus and Fath, 2019).

While I agree that any loss of discrete human life is a tragedy at the level of individual rational human being, sustainability in the Universe must be understood at the level of unified life, where death continues in renewed life, as all living beings see their life intricately connected with and dependent on other living creatures existing now or previously born. There is nothing tragic in this order of things where life is arising from dying life. It is as old as observations from the natural world first noted in the Bible, where seeds need to die in order to germinate new life: "unless a grain of wheat falls into the ground and dies, it remains alone; but if it dies, it produces much grain." (John 12:24). In the same way, a new oak can be born from an acorn fallen on the ground, just as a dead person's ideas may live on, embedded in the broader culture that eventually may impact humanity in a major way. This is the real value of sustained life from the perspective of sustainability, in which "life emerges and self-sustains through a type of self-organization such that each participant 'thing' is simultaneously 'doing its own thing' and 'doing its own thing to fit together'" (Fiscus and Fath, 2019). This holistic understanding of sustainability as a "dance of life" which humans do not control but which they can impact through their decisions – and where humans are expected to do their own thing to fit together – speaks to the intrinsic value of life in all forms that they appear in nature, especially when humans do not understand the role that each individual life plays in keeping the integrity of the dance of life. When once asked what rattlesnakes are good for, the conservationist ecologist John Muir famously said: "they are good for themselves, and we need not begrudge them their share of life" (Muir, 1901/1980). Then, what is the goodness of life? It is an attitude distilled in the commandment to live and let live (Sabau, 2010). It is coming to terms with the deep understanding that life needs to be lived, and it needs to be spared and shared, looking for the good of "the other" before we look for our own good (Schweitzer, 1965). Goodness of life is thus relational and difficult to see from the beginning, as the "goodness does not emerge until the relation is processed, or in some way vetted by the system as a whole" (Fiscus and Fath, 2019). The ecologist Aldo Leopold also understood that a new land ethic was needed, one based in the rightness of acting to preserve "the integrity, stability, and beauty of the biotic community" (Leopold, 1949). Leopold was hoping that his land ethic will change "the role of *Homo sapiens* from conqueror of the land-community to plain member and citizen of it" displaying "respect for his fellow-members, and also respect for the community as such" (Leopold, 1949). Unfortunately, it seems that *Homo sapiens*' respect for life, including human life, is continually on the decline, despite the remarkable technological achievements that have increased the quality of life for many. For instance, over the past 50 years, medical treatments and procedures have significantly diminished human mortality rates and expanded life expectancy (Bailey and Tupy, 2020). Industrial agriculture has reduced the "prevalence of under-nourishment" in the world from 37 per cent of the world population in 1969–1971 to 10.8 per cent in 2018 (Bailey and Tupy, 2020). These human wellbeing trends do not threaten to increase uncontrollably the global population, as global fertility rates have fallen from five children per woman in 1960 to 2.43 children today (European Commission, 2018), in most developed countries (except Israel) which experience declining populations, and only a dozen of countries in Africa have fertility rates higher than three children per woman (Paice, 2021). A demographic study predicts that the world population will peak at 8.9 billion in 2060 and will decline to 7.8 billion by 2100 (EC, 2018). What the carrying capacity of the Earth is in terms of human population is a contested issue; the neo-Malthusians (Ehrlich, 1968; Becker, 2013) believe that there are too many of us, while the "optimists" (Kremer, 1993; Simon, 1996; Lam, 2011) believe that more people means more creative brains to eventually solve humanity's greatest challenges, including upcoming ones. Then there are more extreme ideologues who focus against anthropocentrism, such as the utilitarian bioethicist Peter Singer whose "moral philosophy affords only an impoverished and distorted sense of the value of human life and human beings" (Cordner, 2005). In 1975, Singer published the book *Animal Liberation* in which he recommended the extension of the utilitarian principle "the greatest good of the greatest number" to other animals, as, in his opinion, the boundary between human and "animal" is completely arbitrary. Singer believes that human life has little value or no intrinsic value and that abortion, infanticide, euthanasia, and the killing of people with disabilities can be justified philosophically (Singer, 2002). Such dehumanizing philosophies which question the sanctity and dignity of human life seem to have slowly permeated the modern societies in the Global North, allowing a "culture of death" (Smith, 2016) to replace what societies have traditionally celebrated as a "culture of life". In some of these societies, individuals normally horrified by Stalin's or Hitler's atrocities against human life have come to accept as normal new social practices of taking human lives in the name of, for instance, the right to a dignified death. In Canada, national euthanasia in the form of medical assistance in dying (MAID) was legalized in 2016. Since then, more than 30,000 people have been killed by physicians (Achtman, 2022). Debates on whether MAID should be broadened to individuals that choose death outside of terminal illness or unbearable pain are currently ongoing in the Canadian parliament. Such moves may be seen as both immoral and leading to unsustainable practices, just as Canada's immigration policies aggressively hit new target increases to counter low natural birth rates from its existing population. At the other end of the spectrum, the culture of life is undermined by the new trend popular among young people in liberal democracies to question the right to life of new babies by not starting new families, or by taking their own lives through "deaths of despair" (Stone, 2019). Such attitudes are encouraged by the Voluntary Human Extinction Movement (VHEMT) initiated in United States in 1996 by Les Knight of Portland, Oregon. According to the movement's website, "the hopeful alternative to the extinction of millions of species of plants and animals is the voluntary extinction of one species: Homo sapiens... us" (VHEMT, 2023). Similar messages are spread by the Center for Biological Diversity through its monthly e-publication *Pop X* launched in 2010 with the aim to examine "the connection between unsustainable human population, overconsumption and the extinction of plants and animals around the world" (CBD-PopX, 2023). It is ironic that the subtitle of the *Pop X* publication is "Population, Sustainability, Food, and a Future for All". Of course, this "future for all" does not include humans!

Another threat to humanity as we know it is represented by the transhumanist movement which advocates for the transition of humans to the man-machine (accelerated by artificial intelligence leaps). Initiated in 1998 by Oxford University philosopher Nick Bostrom, the transhumanist intellectual movement, with roots in "Enlightenment rationalism", hopes that pharmaceuticals and genetic engineering will transform humanity and will bring about a blissful future for the super-humans they manage to produce (Bostrom, 2005). More than that, as humans will be offered more radical body and brain augmentation in the hope of achieving a better and longer life, they can even aim for immortality (Chang, 2016). Transhumanism can be seen as a form of rebellion against the existing life-support system which, in the name of technological progress, keeps alive the modern myth of the mechanistic human self. Sustainability requires a new manner of thinking about humans, one affirming their life and creativity, while keeping them away from experiments which aim to dilute the core of what it means to be human and replace it with imitation and an elusive quest for singularity, or the notion that computers will eventually become so advanced that they will erase the boundaries between humans and computers. Instead, perhaps we could celebrate the fact that humans are the only species endowed not only with non-algorithmic features such as morality and emotions (Marks, 2022), but also with bodies and minds that feel pain when a loved one dies, something that a computer will never do.

Connectedness/relationships

We have seen that both the natural system and the social system are characterized by ontic openness, defined as uncertainty, unpredictability, indeterminacy, and emergence, as objectively existing features not only of the world surrounding us but also of ourselves and our physical lives (Nielsen et al., 2020). This fact speaks to the intrinsic unity of the two realms, which exist and evolve together, sometimes in predictable ways – but often in unpredictable ways – making the whole look larger than the sum of its parts, and questioning, as arbitrary, the three-pillars concept of sustainable development (Dawe and Ryan, 2003), as well as the "assumption that the protection of the basis of life is not feasible without economic growth" (Ekardt, 2019). At the heart of this intrinsic unity lies the human being who is both a biological and a social being, living, moving, and acting in both realms, and carrying on relationships with nature, or the natural environment, and with other human beings existing now or in the future. The way a human being chooses to carry out these relationships determines whether they are a sustainable person. We will analyze first the human–nature sustainability relations starting from two premises: (1) that nature exists as a complex and evolving entity "that comes into being and exists independently of human thought and action" (Becker, 2012) – this means that "Nature when we are not watching her behaves in the same way as when we are" (Lewis, 1974); and (2) that humans are complex, evolving, rational, emotional, and telos-driven beings, unable to choose excellence of character (creativity, kindness, etc.) unless they intentionally cultivate certain virtues, and sometimes not even then. This is something we know from St Paul: "For I know that good itself does not dwell in me, that is, in my sinful nature. For I have the desire to do what is good, but I cannot carry it out" (Romans 7:18).

The idea that nature must be studied inductively and without any teleological explanations was introduced by philosopher Francis Bacon (1561–1626), one of the first philosophers of science from the Renaissance period. By explaining that "understanding the basic structures of things" is "a means to transforming nature for human purposes" (Gaukroger, 2001), Bacon started a trend which continues to this day, one which sees nature as purposeless and open to be explored and exploited by humans. The idea that science can be used to control nature was reaffirmed in the 1940s by Norbert Wiener, the father of modern engineering, who stated: "The intention and the result of a scientific inquiry is to obtain an understanding and a control of some part of the universe" (Dougherty, 2019). Based on this philosophy of science, the current mainstream science paradigm has created a rift between nature and humans, seen as separate and antagonistic entities, not only when examined in a lab but also in real life. This paradigm, "conceptually aligned with the Darwinian story of life as 'the struggle for existence'" (Fiscus and Fath, 2019), has led to a generalized state of competitive behaviour as a mainstream practice, disseminated through education, technologies, and the media, and eventually becoming the mainstream culture. In the field of economics, the competition for natural resources begun in the 1950s generated then a vivid debate in the United States of America concerning how nature should be used for human purposes. The "conservationists" (Pinchot, 1947; Krutilla, 1967) believed that natural resources should indeed be used, but wisely, while "preservationists" (Muir, 1875/1980; Leopold, 1949) believed that the "biotic communities" in nature should be preserved for the goodness and rightness of preserving life as a whole. The debate resulted in the victory of conservationists which established a fundamental principle of environmental policies still applied today; it is the utilitarian principle "of use, to take every part of the land and its resources and put it to that use in which it will serve the most people" (Banzhaf, 2016). The visible results of these utilitarian environmental policies are loss, degradation, disruptions, "wasting and poisoning [of] the good and beautiful things of the world, which once were called 'divine gifts' and now are called 'natural resources'" (Berry, 2015), in addition to systemic inequality and increased violence. Seeing nature as "the other", to be conquered and used to increase the amount of human-produced material goods and industrial food, is a hubristic oversight of the "sameness" and mutual relationships human beings share with nature. The oversight becomes dangerous when human actions destroy the very life-sustaining system of which humans are a part. It is known that

> [All] living things require 24 chemical elements, and these must cycle from the environment into organisms and back to the environment. Life also requires a flow of energy [...] Although alive, an individual cannot by itself maintain all the necessary chemical cycling or energy flow.
>
> (Keller and Botkin, 2008)

When we start seeing ourselves not as individual selves, detached from the environment we live in (family, neighbourhood, or country, and the nature they are embedded in), but as part of these functional wholes, we begin to understand that "we cannot possibly flourish alone", as we are "part of one another", and that "we create one another" (Berry, 1977). This renewed understanding of humans' role in the life-supporting web could change the paradigm of human–nature interactions and give a new meaning as well as a renewed practice toward the human–environment relationship. Humans are related to nature not only as biological beings, but as emotional beings also, as they can experience the majesty, beauty, creativity, and mystery of nature as a metaphysical reality, or as "fulness" (Taylor, 2007), or as "the something more" that the understanding provided by the materialist and often reductionist language of science is not adequate enough to express (Cooper et al., 2016). At the same time, humans need to acknowledge "the otherness" of human beings and of nature in the human–nature relationship. We will continue to note with delight that birds or dragonflies can continue to naturally fly, while humans still cannot. Humans will continue to think abstract thoughts and make moral judgements in the same way animals cannot. This, I argue, makes humans responsible agents that are accountable for their actions, because they are endowed with reason, conscience, and free will. These very human traits force us to "face up to the way we live our lives and how we relate to each other and to the planet, knowing that our choices do have consequences that go beyond our small bubble of being" (Gleiser, 2021). Becker (2012) also holds that:

> The human–nature relationship is not something that can be defined merely theoretically or determined by (scientific or economic) rationality. It is something in which the human being is necessarily located, and it

can only fully be understood by living within this relationship and developing personal attitudes of attentiveness and openness to the sameness and otherness as well as to the various levels of mutual encounter.

(Becker, 2012)

A proper understanding of humans' role and responsibility to nature will prompt us to have a holistic understanding of the intrinsic value of all living creatures engaged in the dance of life. Using the attributes of their "otherness", humans can study nature holistically – not as dead matter and energy atoms but as complex and dynamic ecosystems characterized by incredible features such as thermodynamic and ontic openness, connectivity, self-organization, directionality, growth and development, disturbance and decay (Nielsen et al., 2020). These features give ecosystems specific roles in maintaining the global web of life which, among others, feeds us through complex food chains. This knowledge and the awareness that we are part of a beautifully functioning web of life, can equip us with a new mindset, one able to develop a narrative of sustainability based in the value and unity of the ecological and the social systems, and in which "life is not random and unexpected, but very much to be expected when the self-organizing tendency of the universe is fully understood and appreciated" (Kauffman, 1995). This new narrative has the potential to change human practices, as well as culture, enabling humans to do the right thing, living with nature and not against it, and acting to build soils, prevent species extinction, produce healthy food, eliminate water and air pollution, stop ocean acidification and global climate disruption, not only for the sake of humans but for the sake of maintaining the integrity of the planetary web of life.

A sustainability ethics framework

We have argued that the three characteristics of sustainability, durability, goodness of life and relevance of relationships exist objectively and prior to any individual interests and desires. They are available for everyone on Earth and no individual can unilaterally deplete them. This is what gives sustainability its objective value and authority over our lives, as these characteristics exhibit a recommended pattern of behaviour for human sustainability users, as a "morality flowing from the nature of things rather than a construction of the human mind" (Mitchell, 1980). "The way the world is informs the way it ought to be" (Rolston, 1991), thus strangely dissolving the is/ought dichotomy when both the facts and values of nature are discovered as properties of the system. This morality goes beyond respect for the evident limits imposed by the laws of nature, and points to a genuine code of sustainability ethics defining both human beings' place in nature and the right way of using the life-support system which has been given to us. The sustainability ethics appears in the form of guiding principles, good and right in themselves, that will allow life on Earth to continue in this pattern indefinitely. Some of these perceived principles are:

- Enjoy what is provided freely by nature to satisfy your needs, but discover and stay within the existing limits;
- Value and preserve all life, for you are neither the author nor the finisher of life;
- Be just: share the life-sustaining system with other humans and other living creatures;
- Be humble and show gratitude, as you do not own the system providing sustainability, but you are its beneficiary;
- Search to discover and comprehend the ordered complexity of sustainability, its purpose, and its inherent limits;
- Diligently do your part in the functioning of the system;
- Share the truth when you discover it, and contribute to building lasting sustainable communities.

These principles are universal principles but not eternal commandments. They are context specific and, to that degree, unpredictable. Based on these principles, we can identify a set of values that define a sustainability ethics, as an ideal of right relationships of humans with nature and with each other, able to secure life on Earth in complete harmony with the life-sustaining system and not working against it. These sustainability-inspired values are freedom (within pre-established limits), equality (among members of the human species), respect for all life (Leopold, 1949; Schweitzer, 1965), justice (in dealing with others), humility (by assuming our role of creature and not creator), and care (for non-human nature, and for weaker beings). These values have been analyzed as transcendental values (Schwartz, 1992; Kenter et al., 2015; Raymond and Kenter, 2016), accepted across cultures "as concepts or beliefs about desirable end states or behaviours, which transcend specific situations, guide selection or evaluation of behaviour and events, and are ordered by relative importance" (Schwartz, 1992). Transcendental values are independent of human preferences. They are "supernatural realities" "recognized" by humans and not invented by them (Eliot, 2015) that constitute the "irreducible, indispensable prerogative, privilege, and patrimony of human civilization itself" (Aeschliman, 2019). They are often socially shared values (Raymond and Kenter, 2016), as "through the physical linkages existing in nature, a social interconnectedness is forced upon us" (Vatn, 2009). Schwartz (1999) identified a universal set of transcendental values operating both at the individual level and at the cultural/societal level, representing "the implicitly or explicitly shared abstract ideas about what is good, right, and desirable in a society". Among these universal values are "freedom", "equality", "a world at peace", "inner harmony", "wisdom", and "mature love" (Schwartz, 1992). All these values would be excellent candidates for a values-based sustainability ethics. They are core values in religious, pre–modern, and Indigenous communities across the world which have included sustainability in their principles of governance. They all accept that a higher power is behind the Earth's life-sustaining system and a source of sacred knowledge and strength.

For instance, the nations of the Haudenosaunee (Iroquois Confederacy), in their Constitution:

> offer thanks to the earth where men dwell, to the streams of water, the pools, the springs and the lakes, to the maize and the fruits, to the medicinal herbs and trees, to the forest trees for their usefulness, to the animals that serve as food and give their pelts for clothing, to the great winds and the lesser winds, to the Thunderers, to the Sun, the mighty warrior, to the moon, to the messengers of the Creator who reveal his wishes and to the Great Creator who dwells in the heavens above, who gives all the things useful to men, and who is the source and the ruler of health and life.
>
> (Constitution of the Iroquis People, n.d.)

Of these core values, only justice as intergenerational justice has been adopted in the sustainable development definition, through its explicit commitment to leave enough natural resources to satisfy the needs of future generations. However, the tool proposed for achieving sustainable development – unlimited economic growth, based on the human-centred philosophy of ever-growing human wants – can only perpetuate human beings' role of reckless planetary managers which has been assumed by humanity in the last seven decades, and which has led to careless and unjust use of nature's provision to the point of harming the Earth's life-support system, while also producing painful social inequalities. While promoting these values is essential for building a culture of sustainability, there is no guarantee that the current mainstream Western culture, imbued with consumerism, secularism, and moral relativity, can arrive at uniformly accepting and adopting these values in order to secure the future of planet Earth. The existing culture of sustainability is based on the epistemic fallacy that considers sustainability not an objective valuable reality, but a "social construct", or "a reflection of how the current and future quality of the environment is subjectively valued by an individual or group" (CPB, 1996). This culture, by purposefully excluding the concepts related to the sacred from the public discourse, and by limiting accepted human knowledge to knowledge produced by rational thought, has drastically reduced the human potential to see reality holistically and larger than the sum of its parts. This narrow view of reality has impaired human efforts to seek realistic and sustainable solutions to the "wicked problems" of unsustainability that humans have created. The process of changing the current unsustainability culture will likely be long and difficult, and will have to involve a transformation of the individual selves. This chapter has aimed to develop a sustainability ethics rooted in human virtues, a normative construct about how a sustainable person ought to be, think, and act or live sustainably while engaging in all types of relations as part of their life. While it is a theoretical construct, this model of a sustainable person's identity is relevant ontologically because each human being must answer this very question in their real-life practice: *How ought I to be?* While a discussion of virtue may not be particularly popular to a mainstream audience which has conveniently transitioned "from Victorian virtues to modern values" in the 20th century (Himmelfarb, 1996), excellence of character and honesty may still be inspiring for many of us, just as some will continue to recoil and disagree with vices such as cynicism, dishonesty, or greediness. This model may inspire this sort of individuals in their efforts to become sustainable persons. The model is built on three premises: that a human is a social, relational, and communicative being; that a human is a rational, conscious, and intentionally active being with a telos, or longing for a good life; that a human can be transformed from an individualist, atomistic, selfish modern being, which only exists as an universal abstraction (Bhaskar, 2012), into a person of real character displaying certain moral excellence characteristics or virtues which are relevant for sustainability. According to Aristotle, humans are not born morally virtuous, but they can become virtuous by controlling their passions and by forming habits of doing "what one ought" through right education and practice. For Aristotle, moral virtue is "a state of character concerned with choice, lying in a mean, i.e. the mean relative to us, this being determined by a rational principle, and by that principle by which the man of practical wisdom would determine it." (Aristotle, 1947). In other words, excellence of character can be achieved by exercising one's intelligence and practical wisdom (*phronesis*) and by getting an education in the virtues. "For what education in the virtues teaches me is that my good as a man is one and the same as the good of those others with whom I am bound up in human community" (MacIntyre, 1981). The golden mean that Aristotle refers to consists in choosing a "virtuous" action which lies between two less virtuous actions or vices. For instance, telling the truth about the beauty of a painting is the golden mean between maliciously debasing the painting's author for lack of gift and carelessly inflating their ego by lying about the real beauty of the painting. In forest management, the golden mean may be considered selective harvesting, the sustainable option between a clearcut of the forest and leaving the forest untouched. In the Christian tradition, Thomas Aquinas proposed four cardinal virtues – namely, justice, fortitude (courage), temperance, and prudence – as strategic elements for building up excellence of character. These virtues are now considered essential to the leadership for sustainability (Chen, 2012). They are also essential for a sustainable organization of social life, as:

> [P]rudence does what is just, while temperance and fortitude protect prudence from inner and outer threat. Temperance and fortitude seek to make a good person, while it is only the specific aim of justice to make a good person a good citizen.
>
> (Floyd, 2006)

To these four cardinal virtues we should add *sapientia* (wisdom) defined as "a genuine rational knowledge of one's own nature and of transcendent metaphysical realities [....] – knowledge which *homo sapiens* is uniquely fitted and obliged to posess and which regulates all our knowledge of physical nature" (Aeschliman, 2019).

There are four types of relationships that are significant for any human being (Figure 5.1): relationships with nature, relationships with other human beings – contemporaries and future generations – and one's relation with the self. It is important to consider first human beings' relation with the self, because if a person is not going to first set their internal house in order, all other relations may be compromised.

To analyze humans' relation with their self, it is useful to start from Roy Bhaskar's three planar conception of the human being. According to Bhaskar, a human being is a stratified self consisting of an *ego* (the sense of a separate identity from other beings) which pushes beings to be selfish, proud, greedy, or uncooperative; an *embodied personality*, which is the actual visible self, with physical, mental, and emotional properties, which is relative and changing with age; and the *real transcendental self*, also called "the ground-state". The self in the ground-state is not separate from other selves, but "is in fact contiguous with and connected to all other selves and other things generally

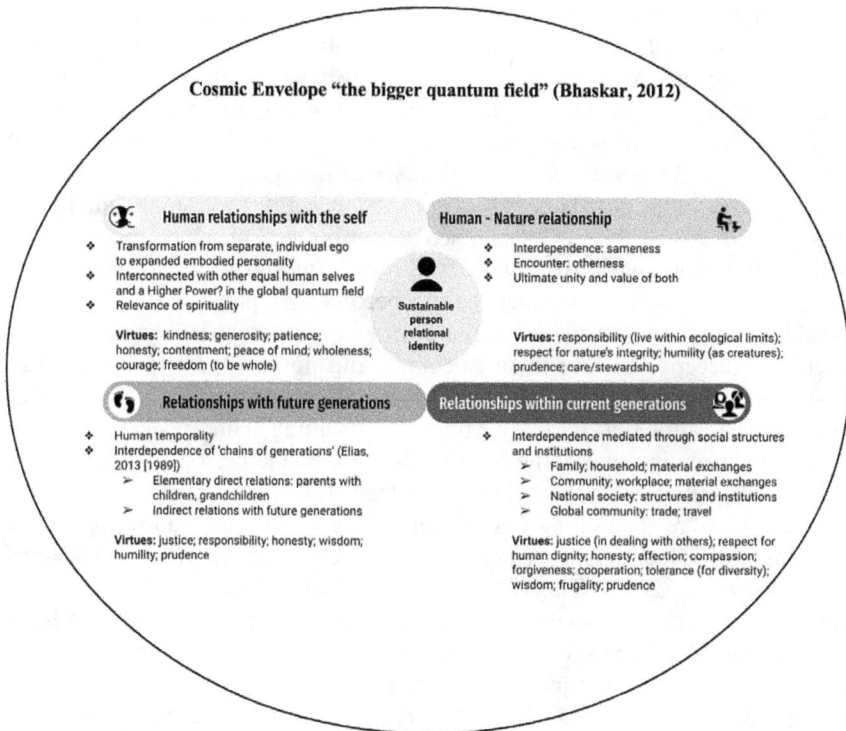

Figure 5.1 Sustainable person's relational identity

and indeed to all creation" (Bhaskar, 2012). Bhaskar calls this interconnected whole, to which all human selves are bound through their conscious and unconscious activities, "the cosmic envelope" or "the bigger quantum field". Bhaskar does not call God the owner of the cosmic envelope, but he insists that we must recognize that a "spirituality within the bounds of secularism" exists, which can refer to a higher power, or a "designing mind with a will" (Meyer, 2021) which is "that source of energy and intelligence, capacity to love and right-action which sustains everything" (Bhaskar, 2012). This higher power, which all faiths acknowledge (Bhaskar, 2012), now starts to be recognized by scientists observing the fine-tuning of the Universe not only to support life but also to secure its continuation in this pattern (Thorvaldsen and Hössjer, 2020; Meyer, 2021). As the human ego is just an illusion, because, in fact, all human beings exist connected to and dependent on other things or beings in the Universe, such as the air they breathe or one's family members and co-workers, it is the ego that must be controlled, minimized, and eliminated. How can this be done? Bhaskar believes that humans can be emancipated from their egos by "shedding the illusion" or the sense of being separate and more important than other human beings. When humans manage to do that, they actually expand the potentialities existing in their true transcendental essential selves, touched by the divine spark which is the source of power over human beings' egoistic tendencies, and they can align their actual embodied self with the transcendental real self. This is a paradox: how can one become stronger by deliberately making oneself weak or vulnerable? This paradox needs to be experienced; it cannot be understood theoretically. The paradox has been mentioned by Donella Meadows in her theory of leverage points: "In the end, it seems that power has less to do with pushing leverage points than it does with strategically, profoundly, madly letting go" (Meadows, 1999). This choice of "letting go" of selfish egoism starts with one's honest self-introspection, and it can amount to a real transformation of the individual self when he/she becomes a "person" (Smith, 2011), capable of transcendentally identifying with other human beings, and consequently to accept and love them as equals, and not use them for selfish purposes, but to treat them, as Immanuel Kant recommended, no longer as means but as aims that have a purpose and meaning (Hill and Zweig, 2003). This transformation marks the birth of the sustainable person and can start a process of setting the person free from fear, guilt, and shame, and free to contribute to "re-enchanting reality from the ground-state, endowing it with meaning and value, the meaning and value that dis-enchantment had drained away" (Bhaskar, 2012). The virtues associated with a sustainable person are kindness, generosity, patience, honesty, contentment, peace of mind, wholeness, courage (as opposed to fear), and freedom to be whole (as opposed to being in bondage to the ego). Being an intentional agent, a sustainable person will reconsider their relations with nature, as well as with other human beings both living now and in the future. When this transformation occurs in numerous individuals, the narrative of the egotistic, atomistic individual that the current modernity-influenced culture has promoted as a universal story will begin to crumble. Sustainable persons will achieve their purpose as interdependent whole human beings interested in the good of other beings, including human beings, and will be made alive to a meaningful life. Such a person will be able to distinguish between liking a person or a thing and truly loving them. For instance, when a sustainable person likes a beautiful flower in the field, they will not kill it by cutting it to satisfy their selfish desire, but they will care for it, watering it and loving it as a part of the landscape.

Human relations with nature start from the acceptance of the ultimate reality and intrinsic value of both nature and the human being. These relations are manifested as both sameness and otherness, and are based on knowledge and respect for nature's otherness and responsibility for how much humans can take from nature and how much care they give nature. The main sustainability virtues in this relationship are responsibility (for living within ecological limits), respect for nature's integrity, prudence, care (stewardship), humility, and restraint. "We should thank God for beer and Burgundy by not drinking too much of them" (Chesterton, 1908).

The relations with other human beings start from the ultimate reality that humans are intentional social beings with goals and aspirations and agency powers, but contingent on other human beings and on the existing structures and institutions of their societies. The social structures and institutions (marriage, family, market, science, church, educational and legal institutions) are relevant for human relationships with their contemporaries, as they shape human behaviours, social rules, and the cultural environment. The main relationships are direct interdependency relations within families, communities, national societies, and the global community. The main sustainability virtues relevant for these relationships are justice (in dealing with others), respect (for human dignity), equality (of all human beings *qua* human beings), honesty, affection, compassion, tolerance, cooperation, wisdom, and prudence. The relations with future generations are prefigured based on the concepts of "human temporality" (Becker, 2012) and "chains of generations" (Elias, 2013/1989), showing that humans are beings in time, and that the current generations are indebted to past generations and will influence future generations through decisions made today. There are two types of relationships with future generations: (1) direct relationships of parents with children and grandchildren – these are interdependence relations characterized by responsibility of older generations to produce new generations, and to pass on to younger generations material goods (wealth, infrastructure, functional nature) and immaterial goods (ecological services, knowledge, wisdom, faith, traditions, confidence and peace), and responsibility of younger generations to provide care, protection, and support to older generations; (2) indirect abstract relations that can be imagined through mental reflection and moral insights. While some may cynically say, "We are always doing something for posterity, but I would fain see posterity do something for us" (Addison,

1721), others consider that "positively influencing the long-term future is a key moral priority of our time" (MacAskill, 2022). The main sustainability virtues relevant for humans' relationship with future generations are justice, responsibility, honesty, wisdom, and humility. As justice appears to be essential for sustainability of human relations with other humans in the current and future generations, I will dig deeper into the meaning and importance of justice for sustainability in the next chapter.

This ethical framework will only thrive if humans are willing to accept that reality exists objectively, and turn away from living in self-made realities, including those attempted to be created in the metaverse, which aim to create platform lock-in or market-based dependencies with respect to our emotional relationships. This includes accepting that humans are "fearfully and wonderfully made", with harmony between body, mind, and soul that function sustainably, according to pre-established rules. One of these rules is that all humans have a longing for eternity in their hearts which cannot come from matter and energy atoms. Satisfying this longing by letting go of selfish desires and illusionary independence will help humans achieve their ultimate goal as human beings. Tinkering with this fine-tuned balance, while it temporarily may bring new happiness, will likely only lead to unhappy, "hybrid" beings in the long run.

While sustainability exists objectively as an intrinsic value, the human response to it has not been unified and we cannot expect a unified response given human nature and humans' freedom to choose. How people respond to sustainability depends on their "preanalytic vision" (Schumpeter, 1954), which is "largely formed based on transcendental values, and relate to themes such as how we judge different states of the world, notions of progress and what we conceptualize as 'good'" (Horcea-Milcu et al., 2019). The current majority in the liberal Global North democracies, under the influence of the modernity worldview shaped by the European Enlightenment movements which modelled our world from the seventeenth century on, have chosen to ignore sustainability as a gift "from beyond" (Cooper et al., 2016) and instead pursue human defined "goods" of "health, wealth and pleasure" (Reno, 2020) as their main source of wellbeing. As I have tried to show in this chapter, the modern self, when centred on criterionless human preferences, is unable to make sustainable choices in our morally fragmented, "after-virtue" societies (MacIntyre, 1981). In these societies, a virtue is defined, ever since David Hume's time, not as excellence of character but as a passion, or "a disposition or sentiment which will produce in us obedience to certain rules" (MacIntyre, 1981). Yet, prior agreement on what the relevant rules might be seems impossible to achieve in the current make-up of our Global North individualistic societies (MacIntyre, 1981). I believe that the Western societies are not completely devoid of virtues; rather, there is a dangerous confusion about virtues that Chesterton very well identified in his book *Orthodoxy:*

> The modern world is full of the old Christian virtues gone mad. The virtues have gone mad because they have been isolated from each other and

are wandering alone. Thus some scientists care for truth; and their truth is pitiless. Thus some humanitarians only care for pity; and their pity (I am sorry to say) is often untruthful.

(Chesterton, 1908)

Despite it being held as a minority view, it is possible to accept that nature's life-sustaining system is a gift and that its proper use can be found by those so inclined to accept it. Empirical studies confirm that people holding transcendental values are prone to engage in environmental stewardship and conservation. Working in Australia, the Solomon Islands, and in the United Kingdom, Raymond and Kenter (2016) showed the importance of assessing "TVs [transcendental values] in ecosystem management and valuation, given that they directly affect intentions to manage ecosystem services and WTP [willingness to pay] for ecosystem services. They also indirectly influence conservation behaviour and monetary values through beliefs and norms."

In conclusion, this is the tragedy of our unsustainable world which lacks a consensus on what sustainability is and what really matters in human lives. While we cannot expect that a global consensus on these important issues will soon appear, we can hope that more people with restless hearts, tired of the existential loneliness they experience in our metaphysically barren world, will choose to do the right thing by "letting go" of their selfish ambitions and desires and gratefully accept the gift of sustainability. It would be an acknowledgement that we are all "basically dependent beings: one upon another, and each on a world that is not of our making" (Crawford, 2009).

This chapter has shown that sustainability is a structured and functional reality with intrinsic value, manifested under three aspects: as persistence, durability, or temporal and spatial continuity of the life-support system; as goodness of life as a whole; and as unity of the biotic community, or the relational character of sustainability. As such, sustainability should be a non-negotiable normative goal overarching all human decisions. However, sustainability is ignored by most individuals in the liberal democracies that function according to the modernist narrative built around an artificial, fragmented, individualistic, and self-interested human self, unable to conceive of and live a sustainable life. The chapter has also discussed the moral values associated with sustainability, and has developed a virtues-based ethical framework of a sustainable person's identity, and the transformation process through which a modern egotistic individual can become a sustainable person.

References

Achtman, A. (2022) "Canada's Orwellian Euthanasia Regime." Law & Liberty. https://lawliberty.org/canadas-orwellian-euthanasia-regime.

Addison, J. (1721) "The Works of the Right Honourable Joseph Addison, Esq." Volume 4 of 4, The Spectator, Number 583, Issue Year: 1714, Issue Date: "Friday, August 20", Printed for Jacob Tonson at Shakespear's Head, London.

Aeschliman, M.D. (2019) *The Restoration of Man. C.S Lewis and the Continuing Case Against Scientism.* Seattle: Discovery Institute Press.

Anderson, P.W. and Stein, DL. (1987) "Broken symmetry, emergent properties, dissipative structures, life: are they related?" In *Self-Organizing Systems: The Emergence of Order,* edited by F.E. Yates, A. Garfinkel, D.O. Walter, and G.B. Yates. New York: Plenum Press.

Archer, M., Decoteau, C., Gorski, P., Little, D., Porpora, D., Rutzou, T., Smith, C., Steinmetz, G., and Vandenberghe, F. (2016) "What is Critical Realism?" *Perspectives: A Newsletter from ASA Theory Section* 39: 4.

Aristotle (1947) "Nicomachean Ethics." In *Man and Man: The Social Philosophers,* edited by S. Commins and R.N. Linscott. New York: Random House.

Bailey, R. and Tupy, M.L. (2020) *Ten Global Trends Every Smart Person Should Know: And Many Others You Will Find Interesting.* Washington, DC: Cato Institute.

Banzhaf, H.S. (2016) "The Environmental Turn in Natural Resource Economics: John Krutilla and 'Conservation Reconsidered'." Resources for the Future, Discussion Paper 16-27: 1–23. Washington, DC.

Bar-On, Y.M., Phillips, R., and Milo, R. (2018) "The biomass distribution on Earth." *PNAS,* 115(25): 6506–6511.

Barrow, J.D. and Tipler, F.J. (1986) *The Anthropic Cosmological Principle.* Oxford: Clarendon Press; New York: Oxford University Press.

Barry, C. (2019) "Charles Taylor on Ethics and Liberty." *Eidos: A Journal for Philosophy of Culture,* 3(9): 83–102. https://doi.org/10.14394/eidos.jpc.2019.0032.

Becker, B. (1997) "Sustainability assessment. A review of values, concepts, and methodological approaches." *Issues in Agriculture,* 10. Washington: The Consultative Group on International Agricultural Research, World Bank.

Becker, C.U. (2012) *Sustainability Ethics and Sustainability Research.* Dordrecht: Springer.

Becker, S. (2013) "Has the world really survived the population bomb? (Commentary on 'how the world survived the population bomb: lessons from 50 years of extraordinary demographic history')." *Demography,* 50(6): 2173–2181. https://doi.org/10.1007/s13524-013-0236-y.

Behe, M.J. (1996) *Darwin's Black Box: The Biochemical Challenge to Evolution.* New York: The Free Press.

Berry, W. (1977) *The Unsettling of America: Culture and Agriculture.* Berkeley, CA: Counterpoint.

Berry, W. (2015) *Our Only World.* Berkeley, CA: Counterpoint.

Bhaskar, R. (2012) *Reflections on MetaReality: Transcendence, Emancipation and Everyday Life.* London and New York: Routledge.

Bostrom, N. (2005) "A history of transhumanist thought." *Journal of Evolution and Technology,* 14(1): 1–25.

Callicott, B. (1989) *In Defense of the Land Ethic: Essays in Environmental Philosophy.* Albany, NY: State University Press of New York Press.

CBD-PopX (2023) "PopX: A newsletter on unsustainable human population growth and the extinction crisis." www.biologicaldiversity.org/programs/population_and_sustainability/pop_x.

Chang, L. (2016) "Want to live forever? Ray Kurzweil Thinks That Might Be Possible Very Soon." *Digital Trends,* March 27, 2016. www.digitaltrends.com/health-fitness/ray-kurzweil-immortality.

Chen, B. (2012) "Moral and Ethical Foundations for Sustainability: A Multidisciplinary Approach." *Journal of Global Citizenship & Equity Education,* 2(2): 1–20.

Chesterton, G.K. (1908) "Orthodoxy." In *G.K. Chesterton Collected Works*, Vol. I, edited by D. Dooley (1986). San Francisco: Ignatius Press.

Constitution of the Iroquis People (n.d.) www.indigenouspeople.net/iroqcon.htm.

Cooper, N., Brady, E., Steen H., and Bryce, R. (2016) "Aesthetic and spiritual values of ecosystems: Recognizing the ontological and axiological plurality of cultural ecosystem 'services'." *Ecosystem Services*, 21: 218–229.

Cordner, C. (2005) "Life and Death Matters: Losing a Sense of the Value of Human Beings." *Theoretical Medicine and Bioethics*, 26: 207–226.

CPB (1996) *Economie en milieu: op zoek naar duurzaamheid*. Centraal Plan Bureau. The Hague: SDU Publishers.

Crawford, M.B. (2009) *Shop Class as Soulcraft: An Inquiry into the Value of Work*. New York: Penguin Group.

Dawe, N.K. and Ryan, K.L. (2003) "The Faulty Three-Legged-Stool Model of Sustainable Development." *Conservation Biology*, 17(5): 1458–1460.

Dembski, W.A. (1998) *The Design Inference: Eliminating Chance Through Small Probabilities*. Cambridge: Cambridge University Press.

Dodd, M.S., Papineau, D., Grenne, T.*et al.* (2017) "Evidence for early life in Earth's oldest hydrothermal vent precipitates." *Nature*, 543(7643): 60–64. https://doi.org/10.1038/nature21377.

Dougherty, E.R. (2019) "The Decline of American Science and Engineering." *American Affairs*, III(1): 113–126.

Ehrenfeld, J. (2008) *Sustainability by Design: A Subversive Strategy for Transforming Our Consumer Culture*. New Haven and London: Yale University Press.

Ehrlich P. (1968) *The Population Bomb*. New York: Ballantine Books.

Ekardt, F. (2019) *Sustainability Transformation, Governance, Ethics, Law*. Cham, Switzerland: Springer.

Elias, N. (2013) *Studies on the Germans*. Dublin: University College Dublin Press. (Original work published 1989)

Eliot, T.S. (2015) "Donne in Our Time." *The Complete Prose of T.S. Eliot: The Critical Edition*, edited by J. Harding and R. Schuchard, Vol. 4, *English Lion, 1930–1933*. Baltimore: Johns Hopkins University Press, pp. 369–382.

European Commission (EC) (2018) *Demographic and Human Capital Scenarios for the 21st Century: 2018 assessment for 201 countries*, edited by W. Lutz, A. Goujon, K. C. Samir, M. Stonawski, N. Stilianakis. Luxembourg: Publications Office of the European Union.

Fiscus, D.A. and Fath, B.D. (2019) *Foundations for Sustainability. A Coherent Framework of Life-Environment Relations*. London: Elsevier Academic Press.

Floyd, S. (2006) "Aquinas' moral philosophy." The Internet Encyclopedia of Philosophy. www.iep.utm.edu/a/aq-moral.htm.

Gale, F.P. (2018) *The Political Economy of Sustainability*. Cheltenham, UK and Northampton, MA: Edward Elgar.

Gaukroger, S. (2001) *Francis Bacon and the Transformation of Early-Modern Philosophy*, Cambridge: Cambridge University Press.

Gewirth, A. (1973–1974) "The 'Is-Ought' Problem Resolved." *Proceedings and Addresses of the American Philosophical Association*, 47: 34–61.

Gleiser, M. (2021) "Do the laws of physics and neuroscience disprove free will?" *Big Think*, 13(8). https://bigthink.com/13-8/physics-neuroscience-free-will.

Haudenosaunee Confederacy (2020) "Values." www.haudenosauneeconfederacy.com/values.

Henning, B.G. (2006) "Is There an Ethics of Creativity?" *Chromatikon*, 2: 161–173.

Hill, T. and Zweig, A. (2003) *Groundwork for the Metaphysics of Morals.* New York: Oxford University Press.

Himmelfarb, G. (1996) *The De-Moralization of Society: From Victorian Virtues to Modern Values.* New York: Vintage Books.

Horcea-Milcu, A.-I., Abson, D.J., Apetrei, C.I., Duse, I.A., Freeth, R., Riechers, M., Lam, D.P.M., Dorninger, C., and Lang, D.J. (2019) "Values in transformational sustainability science: Four perspectives for change." *Sustainability Science.* https://doi.org/10.1007/s11625-019-00656-1.

Hume, D. (1978) *A Treatise of Human Nature*, edited by L.A. Selby-Bigge, 2nd ed. rev. Oxford: Clarendon Press. (Original work published 1888)

Hursthouse, R. (1999) *On Virtue Ethics.* New York: Oxford University Press.

IPBES (2019) *Global Assessment Report on Biodiversity and Ecosystem Services*, edited by E.S. Brondizio, J. Settele, S. Díaz, and H.T. Ngo. Bonn: IPBES Secretariat. https://doi.org/10.5281/zenodo.3831673.

Kauffman, S. (1995) *At Home in the Universe: The Search for the Laws of Self-Organization and Complexity.* New York: Oxford University Press.

Keller, E.A. and Botkin, D.B. (2008) *Essential Environmental Science.* Hoboken, NJ: Wiley.

Kenter, J.O., O'Brien, L., Hockley, N., Ravenscroft, N., Fazey, I., Irvine, K.N., Reed, M.S., Christie M., Brady, E., Bryce, R., Church A., Cooper, N., Davies, A., Evely, A., Everard, M., Fish, R., Fisher, J.A., Jobstvogt, N., Molloy, C., Orchard-Webb, J., Ranger, S., Ryan, M., Watson, V., Williams, S. (2015) "What are shared and social values of ecosystems?" *Ecological Economics*, 111: 86–99.

Kibert, C.J., Tiele, L., Peterson, A., Monroe, M. (2012) *The Ethics of Sustainability.* http://rio20.net/wp-content/uploads/2012/01/Ethics-of-Sustainability-Textbook.pdf.

Kohak, E. (2000) *The Green Halo: A Bird's-Eye View of Ecological Ethics.* Chicago: Open Court.

Kremer, M. (1993) "Population Growth and Technological Change: One Million B.C. to 1990." *The Quarterly Journal of Economics*, 108(3): 681–716.

Krutilla, J.V. (1967) "Conservation Reconsidered." *American Economic Review*, 57(4): 777–786.

Lam, D. (2011) "How the World Survived the Population Bomb: Lessons from 50 Years of Extraordinary Demographic History." *Demography*, 48: 1231–1262.

Leopold, A. (1949) *A Sand County Almanac: and Sketches Here and There.* New York: Oxford University Press.

Lewis, C.S. (1974) *Miracles: A Preliminary Study.* New York: HarperOne.

Lewis, G.F. and Barnes, L.A. (2016) *A Fortunate Universe: Life in a Finely Tuned Cosmos.* Cambridge: Cambridge University Press.

Lineweaver, C.H. and Egan, C.A. (2008) "Life, Gravity and the Second Law of Thermodynamics." *Physics of Life Reviews*, 5: 225–242.

MacAskill, W. (2022) *What We Owe the Future.* New York: Basic Books.

MacIntyre, A.C. (1981) *After Virtue: A Study in Moral Theory.* Notre Dame, IN: University of Notre Dame Press.

Marks, R.J. (2022) *Non-Computable You. What You Do that Artificial Intelligence Never Will.* Seattle: Discovery Institute Press.

Martin, M. and Augustine, K. (eds.) (2015) *The Myth of an Afterlife: The Case against Life after Death.* Lanham, MD: Rowman and Littlefield.

Masaeli, M. and Sneller, R. (eds) (2018) *Cosmic Consciousness and Human Excellence: Implications for Global Ethics.* Newcastle upon Tyne, UK: Cambridge Scholars Publishing.

McGrath, S.J. (2023) Personal Communication.

Meadows, D. (1999) *Leverage Points: Places to Intervene in a System.* Hartland, VT: Sustainability Institute.

Meyer, S.C. (2009) *Signature in the Cell DNA and the Evidence for Intelligent Design.* New York: HarperOne.

Meyer, S.C. (2021) *Return of the God Hypothesis: Three Scientific Discoveries That Reveal the Mind Behind the Universe.* New York: HarperCollins.

Millennium Ecosystem Assessment (2005) *Ecosystems and Human Well-Being: Synthesis.* Washington, DC: Island Press.

Mitchell, B. (1980) *Morality: Religious and Secular.* Oxford: Oxford University Press.

Muir, J. (1980) "Wild Wool." In *Wilderness Essays*, edited by F. Buske. Salt Lake City: Peregrine Smith, pp. 227–242. (Original work published 1875)

Muir, J. (1901) 1980. "The Yellowstone National Park." In *Wilderness Essays*, edited by F. Buske. Salt Lake City: Peregrine Smith, pp. 178–219. (Original work published 1901)

Murphy, Jr., T.W., Murphy, D.J., Love, T.F., LeHew, M.L.A., and McCall, B.J. (2021) "Modernity is incompatible with planetary limits: Developing a PLAN for the future." *Energy Research and Social Science*, 81. https://doi.org/10.1016/j.erss.2021.102239.

Nielsen, S.N., Fath, B.D., Bastianoni, S., Marques, J.C., Muller, F., Patten, B. C., Ulanowicz, R.E., Jorgensen, S.E., and Tiezzi, E. (2020) *A New Ecology Systems Perspective* (2nd edition). Amsterdam, Cambridge, MA: Elsevier.

Norton, B.G. (1996) "Integration or Reduction: Two Approaches to Environmental Values." In *Environmental Pragmatism*, edited by A. Light and E. Katz. Abingdon and New York: Routledge.

Paice, E. (2021) *Youthquake: Why African Demography Should Matter to the World.* London: Head of Zeus.

Pinchot, G. (1947) *Breaking New Ground.* New York: Harcourt, Brace & Co.

Prigogine, I. (1980) *From Being to Becoming. Time and Complexity in the Physical Sciences.* San Francisco, CA: W.H. Freeman and Company.

Raymond, C.M., and Kenter, J.O. (2016) "Transcendental values and the valuation of ecosystem services." *Ecosystem Services*, 21: 241–257.

Reno, R.R. (2020) "Against Indifferentism." *Law and Liberty Forum*, November 4, 2020. https://lawliberty.org/forum/against-indifferentism/?utm_source=LAL+Updates&utm_campaign=e4d9624e91-LAL_Daily_Updates&utm_medium=email&utm_term=0_53ee3e1605-e4d9624e91-72470665.

Rescher, N. (2008) "Moral Objectivity." *Social Philosophy and Policy*, 25(1): 393–409.

Richter, D., Grün, R., Joannes-Boyau, R., Steele, T.E., Amani, F., Rué, M., Fernandes, P., Raynal, J.-P., Geraads, D., Ben-Ncer, A., Hublin, J.-J., and McPherro, S. P. (2017) "The age of the hominin fossils from Jebel Irhoud, Morocco, and the origins of the Middle Stone Age." *Nature*, 546: 29396.

Rockström, J., Steffen, W., Noone, K., Persson, Å., Chapin, F.S.III, Lambin, E., Lenton, T.M., Scheffer, M., Folke, C., Schellnhuber, H., Nykvist, B., De Wit, C.A., Hughes, T., van der Leeuw, S., Rodhe, H., Sörlin, S., Snyder, P.K., Costanza, R., Svedin, U., Falkenmark, M., Karlberg, L., Corell, R.W., Fabry, V.J., Hansen, J., Walker, B., Liverman, D., Richardson, K., Crutzen, P., and J. Foley, J. (2009) "Planetary boundaries:exploring the safe operating space for humanity." *Ecology and Society*, 14(2): 32. http://www.ecologyandsociety.org/vol14/iss2/art32.

Rolston, H. III (1986) *Philosophy Gone Wild: Essays in Environmental Ethics.* Amherst, NY: Prometheus.

Rolston, H. III (1988) *Environmental Ethics.* Philadelphia: Temple University Press.

Rolston, H. III (1991) "Environmental Ethics: Values In and Duties to the Natural World." In *Ecology, Economics, Ethics: The Broken Circle*, edited by F.H. Bormann and S.R. Kellert. New Haven: Yale University Press, pp. 73–96.

Rosen, R. (1991) *Life Itself: A Comprehensive Inquiry into the Nature, Origin, and Fabrication of Life.* New York: Columbia University Press.

Rosen, R. (2000) *Essays on Life Itself.* New York: Columbia University Press.

Sabau, G. (2010) "Know, Live, and Let Live: Towards a Redefinition of the Knowledge-Based Economy – Sustainable Development Nexus." *Ecological Economics*, 69: 1193–1201.

Sagan, C. (2022) "Biography." Encyclopaedia Britannica online. www.britannica.com/contributor/Carl-Sagan/2564.

Schumpeter, J.A. (1954) *History of Economic Analysis.* New York: Oxford University Press.

Schwartz, S.H. (1992) "Universals in the content and structure of values: theoretical advances and empirical tests in 20 countries." *Advances in Experimental Social Psychology*, 25: 1–65.

Schwartz, S.H. (1999) "A theory of cultural values and some implications for work." *Applied Psychology*, 48: 23–247.

Schweitzer, A. (1936) "The Ethics of Reverence for Life." *Christendom*, 1(2): 225–239.

Schweitzer, A. (1965) *The Teaching of Reverence for Life.* New York: Holt, Rinehart and Winston.

Schweitzer, A. (1987) *The Philosophy of Civilization.* Vol. 1, *The Decay and the Restoration of Civilization.* Vol. 2, *Civilization and Ethics*, translated by C.T. Campion. Buffalo, NY: Prometheus Books.

Simon, J. (1996) *The Ultimate Resource* 2. Princeton, NJ: Princeton University Press.

Singer, P. (2002) *Unsanctifying Human Life.* Wiley-Blackwell.

Smith, C. (2011) *What Is a Person? Rethinking Humanity, Social Life, and the Moral Good from the Person Up.* Chicago: University of Chicago Press.

Smith, W.J. (2016) *Culture of Death: The Age of "Do Harm" Medicine.* New York/London: Encounter Books.

Steffen, W., Richardson, K., Rockström, J., Cornell, S.E., Fetzer, I., Bennett, E.M., Biggs, R., Carpenter, S.R., de Vries, W., de Wit, C.A., Folke, C., Gerten, D., Heinke, J., Mace, G.M., Persson, L.M., Ramanathan, V., Reyers, B., and Srlin, S. (2015) "Planetary boundaries: Guiding human development on a changing planet." *Science*, 347(6223): 1–10.

Stone, L. (2019) *Red, White, and Gray Population Aging, Deaths of Despair, and the Institutional Stagnation of America.* American Enterprise Institute. www.aei.org/wp-content/uploads/2019/06/Red-White-and-Gray.pdf.

Taylor, C. (1991) *The Malaise of Modernity.* Concord, ON: Anansi.

Taylor, C. (2007) *A Secular Age.* Cambridge, MA: Cambridge University Press.

Taylor, P.W. (1981) "The Ethics of Respect for Nature." *Environmental Ethics*, 3: 197–218.

Thomson W. (1862) "On the age of the sun's heat." *Macmillan's Magazine*, 5: 288–293, 394–368.

Thorvaldsen, S. and Hössjer, O. (2020) "Using statistical methods to model the fine-tuning of molecular machines and systems." *Journal of Theoretical Biology*, 501, 1–14.

United Nations (UN) (1992) *Rio Declaration on Environment and Development*. New York: United Nations.

United Nations (UN) (2002) "Johannesburg Declaration on Sustainable Development." In *Report of the World Summit on Sustainable Development*. New York: United Nations, pp. 1–5.

UN-CBD (2022) "Conference of the Parties to the Convention on Biological Diversity Fifteenth Meeting–Montreal, Canada, 7–19 December 2022." www.cbd.int/doc/c/e6d3/cd1d/daf663719a03902a9b116c34/cop-15-l-25-en.pdf.

Vatn, A. (2009) "An institutional analysis of methods for environmental appraisal." *Ecological Economics*, 68: 2207–2215.

VHEMT (2003) The Voluntary Human Extinction Movement. www.vhemt.org.

Vitousek, P.M., Mooney, H.A., Lubchenco, J., and Melillo, J.M. (1997) "Human Domination of Earth's Ecosystems." *Science*, 277: 494–499.

Wackernagel, M. and Rees, W. (1996) *Our Ecological Footprint: Reducing Human Impact on the Earth*. Philadelphia: New Society Publishers.

World Commission on Environment and Development (WCED) (1987) *Our Common Future*. Oxford: Oxford University Press.

6 Sustainability as justice

Beyond distributive justice

We have seen that sustainability, the ability of planet Earth to persist and provide a lasting life-support system for all living things, exists objectively, wired in the Earth's natural systems (Hueting and Reijnders, 1998). It has visible aspects (air, water, trees, fish) and invisible aspects, such as entropy, photosynthesis, carbon cycling, and protein folding, which are freely and indiscriminately available for all living beings. Thus, a basic principle of justice is embedded in sustainability, implying an appropriate or morally obliging pattern of distribution of benefits and burdens provided by the planetary natural systems to all living inhabitants of the Earth, now and in the future. This principle of justice makes clear that all living creatures may enjoy freely, as a gift, fresh water, sunshine, dewy mornings, and clean air, and none is protected from the potential destruction of naturally occurring events such as earthquakes, floods, or volcanic eruptions. We would expect that rational human beings, interested in the preservation and flourishing of human life, would organize their aggregate existence around a just and careful use of nature's resources, one able to preserve the structural and functional integrity of natural systems. Yet, we have seen from the history of economic development in the twentieth century that this has not been the case. An increase in human ingenuity in science and technological innovation has led to human conquest of nature and appropriation of more and more of nature's beneficence for the satisfaction of human wants, as "man, unlike other creatures, is gifted and cursed with an imagination which extends his appetites beyond the requirements of subsistence" (Niebuhr, 1960). The injustice against nature has been accompanied by injustice in human societies organized around satisfaction of subjective consumer preferences, based on market allocation of resources and a specific distributive justice system which has resulted in the current unsustainable societies threatened by climate change and rampant inequalities and poverty (Gough, 2017; Piketty, 2014). In 1987, the UN Brundtland report entitled *Our Common Future* included the principle of intergenerational justice based on human need satisfaction in the concept of "sustainable development" defined as "meeting the needs of the present generations, without jeopardizing the abilities of future generations to meet their own needs" (WCED, 1987). The report specifically mentions that depleting

DOI: 10.4324/9781003307587-7

natural capital and exporting anthropogenic environmental risks to future generations is unacceptable. However, the practical application of the "sustainable development" concept over the last three decades has led humanity on an unsustainable path, manifested in deep and protracted environmental and social crises, while "only limited and halting steps are being taken to secure *Homo Sapiens* a liveable future" (Gale, 2018). A 2022 assessment of the political impact of the global Sustainable Development Goals (SDG) adopted in 2015 by the United Nations (UN) organization notes that the goals have had only "limited transformative impact" and, overall, did not shift the world on to a sustainable and resilient path: in fact, "we are far away from 'free[ing] the human race from the tyranny of poverty and want and heal [ing] and secur[ing] our planet'", as the UN general assembly agreed in 2015 (UNGA 2015: preamble) (Biermann et al., 2022). For instance, in 2020, 160 million children – some as young as five – were involved in child labour globally, accounting for almost one in ten children worldwide whose right to childhood has been violated (ILO-UNICEF, 2022), in spite of the UN Convention on the Rights of the Child adopted in 1989.

This chapter aims to identify why countries in the Global North have become unsustainable, and to analyze the conditions for achieving just and sustainable societies, defined not only as those societies which have solved the difficult "problem of the equitable distribution of the physical and cultural goods which provide for the preservation and fulfillment of human life" (Niebuhr, 1960), but as societies with free individuals acting justly in their relations with nature and with other human beings because this is the right thing to do. Writing in 1932, Reinhold Niebuhr warned about the difficulty of achieving a just and sustainable social organization. He noted that while a human individual might act morally and consider the interests of others above their own interest, when organized in groups or societies, humans tend to act immorally guided by a "collective egoism, compounded of the egoistic impulses of individuals" (Niebuhr, 1960). Building on Niebuhr's assessment, two additional reasons stand out as to why Global North societies have skewed toward unsustainability. The first is an inability to look at society not as an artificial aggregation of individualistic interests but as "a partnership not only between those who are living, but between those who are living, those who are dead, and those who are to be born" (Burke, 1910). The second reason is a lack of agreement in our postmodern societies on a unified conception of justice, as "the necessary basis for political community" (MacIntyre, 1981). Two thousand and fifty years ago, Plato and Aristotle defined justice as the main virtue making the political life of a community possible, as "justice, alone of the virtues, is thought to be 'another's good' (Plato, 1947) because it is related to our neighbour, for it does what is advantageous to another, either a ruler, or copartner" (Aristotle, 1947). This chapter will first critically analyze why the liberal democratic countries of the Global North have become increasingly unsustainable when they chose to organize their societies around continuous economic growth, based on global competition in free markets (economic capitalism), but lowered the institutional and social protection extended to both nature and human resources (pre-capitalist social order), the latter being left to be accomplished by John Rawls' minimalist "principles of social justice for a well-ordered society" (Rawls, 1971). Next, I will propose a normative universal theory of sustainability as justice (primary and reactive), rooted in the inalienable human right to freedom and "elementary preconditions of freedom", such as life, health, subsistence (food, water), security, climate stability, education, absence of war, etc. (Ekardt, 2020).

The road to unsustainability

A principle of intergenerational justice was inferred in the sustainability concept developed by classical economists starting in the nineteenth century. For instance, J.S. Mill considered the possibility of a "stationary state" for the economy, an economy with constant capital and population stocks which does not grow physically but is able to improve humans' "Art of Living" and bring about "better mental culture, and moral and social progress [.....] when minds cease to be engrossed by the art of getting on" (Mill, 1986). A "stationary economy" would be possible in a classical liberal economy based on principles of good governance and liberty for individual humans, where growth in markets would be controlled by the state – acting as a "night watchman state that would set the boundaries for the natural growth of the market, like a shepherd tending his flock", and where people "needed to be nurtured to first find themselves, in order to act as legitimate citizens in a liberal society" (Mirowski, 2014). In his book *Principles of Political Economy* (1848), J.S. Mill showed concern for future generations when he stated:

> If the earth must lose that great portion of its pleasantness which it owes to things that the unlimited increase of wealth and population would extirpate from it, for the mere purpose of enabling it to support a larger, but not a happier or better population, I sincerely hope, for the sake of posterity, that they will be content to be stationary, long before necessity compels them to it.
>
> (Mill, 1986)

Mill's "stationary economy" concept was further developed in the twentieth century by ecological economists (Daly, 1973; Goodland, 1995, Czech, 2002) under the name of the "steady-state economy". A "steady-state economy" is a sustainable economy which develops within planetary ecological limits, and aims to "maximize current welfare subject to the constraint that natural capital be maintained over generations" (Daly, 1998). In 1946, Sir John Hicks, in his book *Value and Capital*, introduced the concept of a "progressive economy" (as opposed to J.S. Mill's 1848 "stationary economy"), out of concern for *future* generations. He said: "In the process of economic development, the amount of capital per unit of labour should not decrease, because this would diminish the growth potential of future generations" (Hicks, 1950). Out of concern for *current* generations, Hicks introduced the concept of "sustainable income", defined as "a man's income as the maximum value which he can consume during a week, and still expect to be as well off at the end of the week as he was at the beginning" (Hicks, 1950). Both of Hicks' definitions denote prudential concern, and can serve as a guide for sustainable production and consumption in any society interested in sustainable development. The classical liberal presumption that the growth in markets will be controlled by the national states was challenged during the Great Depression's (1929–1939) worldwide downturn in economic activity, which in the post-World War era has led to an increase in the role of national governments in managing the macro elements of the economy. According to J.M. Keynes' *The General Theory of Employment, Interest and Money* (Keynes, 1936/1965), the government's role was to prevent another depression by managing the aggregate demand through discretionary fiscal policies aiming to secure full employment and to provide basic economic security ("welfare") to its citizens. A variant of neo-Keynesian policy, advocating no budgetary constraints on government spending as long as the government is the issuer of money, was followed by most developed countries as part of the response to the 2008 financial and economic crisis as well as to the global pandemic started in 2019. Fiscal policies based on this theory, dubbed as the Modern Money Theory (Wray, 1998; Mosler, 2013; Kelton, 2020), have led, as of January 2023 to huge government debts in countries such as the United States and Japan, and protracted global inflation. Based on Keynes' macroeconomics theory, welfare states, displaying extensive social security systems, including public education, arose after the Second World War in the countries of Western Europe and North America, providing protection both against economic risks through income and wealth redistribution, and against social risks through free public health care and education systems. These welfare states took different forms in various capitalist countries, being classified according to the underlying political traditions in three groups, as social democracy, Christian democracy, and liberalism (Esping-Andersen, 1990). In these welfare state regimes, the system of income redistribution was largely based on John Rawls' theory of fair distributive justice (Rawls, 1971) which still informs most tax systems in capitalist countries. The "fairness" consists in the government's duty to collect taxes from better-off citizens and to redistribute income to less well-off citizens (the maxi-min principle). A similar principle, called the Kaldor–Hicks "compensation principle" (1939), was defined in economics starting from Pareto's concept of "optimal" functioning of markets. The principle states that social welfare increased if those who gained in the market should potentially be able to compensate those who lost for the loss in their welfare and still be better off.

In 1956, economists Robert Solow and Trevor Swan individually developed a theory of economic growth (Solow, 1956; Swan, 1956), a model which explains how the long-run growth rate of a capitalist economy (expressed as growth in real GDP) depends on investment in human-made capital, population growth, and technological change. The model became very popular (in 1987, Solow received the Economic Sciences Prize – the equivalent of the Nobel Prize – for his theory of economic growth), even if it completely ignored the contribution of natural capital to economic growth, and thus concealed the social environmental costs of continuous economic growth. In a 1991 article, Solow discussed sustainability from a neoclassical economics perspective and defined it "as a general moral obligation" to preserve for future generations "the capacity to be well off, to be as well off as we", and as "a general guide to policies that have to do with investment, conservation and resource use" (Solow, 1991). However, Solow never explained what these policies would be. Yet, in a 1974 article on "Intergenerational Equity and Exhaustible Resources", while exploring the consequences of the "fair" distribution of wealth according to John Rawls' theory of justice, Solow noted that with a growing population and technological progress, maintaining constant consumption per head may not be desirable (Beckerman, 1998). This conclusion likely resonates with most neoclassical economists who believe that sustainability, even in its "weak" version (meaning allowing substitution of one form of capital for another in the capital aggregate), should not be the "optimal" choice or a "constraint" in organizing economic activity, "insofar as society seeks to maximize welfare" (Beckerman, 1998). Solow's model of economic growth continues to be used today by all the countries that aim to grow their gross domestic product within the framework of a capitalist economy, characterized by private property of the means of production, free labour markets (based on wages), free goods and services markets, as well as endless accumulation of capital and growth in production for maximizing profits. History shows that growth-oriented economies have produced, particularly right after the Second World War, significant prosperity, heightened scientific progress, and increases in their citizens' life expectancy. However, studies show that the social and environmental costs of continuous economic growth have also increased and now threaten the very prospects of continuity of human life on Earth (Czech, 2002; McNeill and Engelke, 2014; Chernilo, 2017). A 2015 article discussing the "Great Acceleration", a term referring to the holistic, comprehensive, and interlinked changes of the post-1950 era, shows that profound changes due to intensified human activities are now occurring simultaneously across the socio-economic and biophysical spheres of the Earth system, and are encompassing far more than climate change. The article shows that these changes have taken place mostly in industrialized countries, organized since 1961 in the Organization for Economic Cooperation and Development (OECD): "Most of the population growth since 1950 has been in the non-OECD world but the world's economy (GDP), and hence consumption, is still strongly dominated by the OECD world" (Steffen et al., 2015). The article also shows that there is a positive correlation between intensive economic activity and greenhouse gas pollution, as the industrialized economies have grown since the 1950s by intensively using cheap and dirty fossil fuels. In the 1970s, most industrialized countries established Environmental Protection Agencies aiming to protect people and the environment from environmental risks through environmental regulation. However, the efficiency of environmental policies is questioned in studies showing that "continuing growth of production and consumption in the OECD has precipitated new environmental problems and driven dangerous levels of global warming" (Gough, 2017). Studies also show that continuous economic growth produces winners and losers not only among countries but also within countries, by increasing levels of economic inequality. A 2011 OECD study entitled *Divided We Stand: Why Inequality Keeps Rising* reported that inequality increased in 17 out of 22 OECD countries, as measured by the Gini coefficient, a standard measure of income inequality that ranges from 0 (when everybody has identical incomes) to 1 (when all income goes to only one person). Across the OECD states, the Gini coefficient stood at an average of 0.29 in the mid-1980s and rose by 10 per cent to almost 0.316 by the late 2000s (OECD, 2011). In order to understand why continuous economic growth produced these unsustainable outcomes, we need to study a trend that was building up in the economic and social policies of industrialized capitalist countries since the end of the 1930s. The trend, aiming to restore the supremacy of uncontrolled markets, is called neoliberalism.

Neoliberalism

Neoliberalism arose in the late 1930s as a theoretical response of liberal political economists to three twentieth-century ideologies that advocated large states: *communism* (as the most prominent form of socialism), *fascism,* and *social democracy.* The Austrian economist Friedrick Hayek's 1944 book on markets as informational systems characterized by spontaneous order was a response to communist central planning and to Keynes' macroeconomic intervention policies (Hayek, 1944/2007). Milton Friedman's monetarism (Friedman, 1962/2002) was a response to Keynesian macroeconomic policy, and James Buchanan's public-choice research program (Buchanan, 1979) was a response to the economics of general equilibrium and market failure economics (Vallier, 2022). By the end of the 1970s, certain ideas and practices inspired by neoliberal thinkers became mainstream in capitalist states' economic and social policies. These ideas were questioning the large size of the government (welfare states), denigrating collective action, and proclaiming the superiority of the markets (Gough, 2017). Neoliberal ideas were using political economy concepts to transform the capitalist society into a market society in which individuals were free to act as producers and consumers participating intensively in competitive markets in order to maximize their profits and utility, "under the implicit presumption that 'government' is benevolently despotic" (Buchanan, 1979) and that democracy can be constitutionally limited. By 2017, a global neoliberal agenda was identified "for pushing deregulation on economies around the world, for forcing open national markets to trade and capital, and for demanding that governments shrink themselves via austerity and privatization" (Metcalf, 2017). The privatization of government-owned assets has spread around the world since 1997, when the International Monetary Fund (one of the global lending agencies) "included privatization as a standard condition of its structural adjustment lending" (Davis et al., 2000).

As these agenda items are very different from the principles of classical liberalism capitalism, it is useful to analyze the main neoliberal ideas developed by the three political economists mentioned above to see whether these ideas are the cause of the current ecological, social, and economic unsustainability in the advanced capitalist societies of the Global North.

The preferred institution of neoliberalism is the free market. It is preferred because markets are considered to be morally neutral. They allow people with different value systems to benefit from one another, even as they do not guarantee moral outcomes (i.e. the market's efficient outcomes are not equitable; they can make some people rich and some people poor) (Hayek, 1978). However, in Hayek's view, the markets are super-institutions which have the ability to aggregate different people's preferences and to reveal values held as expressed in market prices, an ability which can impact the economic and social structure of a society, by influencing the behaviour of producers and consumers participating in the market. In neoliberal capitalism, human beings are seen as autonomous selves, as a "modern man seeks not to live with his neighbour, but to become an island, even when his natural setting dictates moral community" (Buchanan, 1979). The ideal neoliberal human being is the rational and bold entrepreneur, who is an agent of change, a promoter of economic growth through "creative destruction" innovation (Schumpeter, 1934). The most important virtue in neoliberalism – more important than justice, or anything else – is freedom, defined "negatively" as "freedom to choose" and, most importantly, defined as the freedom of corporations to act as they please (Mirowski, 2014). In 1962, the monetarist economist Milton Friedman (1962/2002) defended the corporations as providers of social benefits. He said that a corporation's first duty is to maximize profits for shareholders and that public policy should ensure that the maximization of profit works to the benefit of all. However, Friedman did not specify what type of public policies should be used for socially spreading the benefits produced by corporations. In August 2019, a statement of the Business Roundtable of United States' CEOs redefined the purpose of a corporation, noting that a corporation's purpose is not only to maximize profits for shareholders, but also to deliver long-term value to all their stakeholders, namely customers, employees, suppliers, the communities in which they operate, as well as to their shareholders (Business Roundtable, 2019). This initiative has generated a large literature on stakeholder capitalism (Stout, 2012; Kaplan, 2019) and numerous actions for promoting corporate social responsibility (CSR), as well as discussions about the ability of corporations to achieve both profits and environmental, social, and governance (ESG) purposes, given the multitude of competing interests that a corporation will need to satisfy under the stakeholder capitalism regime. For instance, one author believes that if it is difficult to reach an agreement on how to maximize wealth, it would be impossible to agree about how to divide the wealth among so many stakeholders (Bainbridge, 2023).

Neoliberals believe in the basic principles of democracy, such as democratic rights to equal participation in elections and in voting, and protection of parliamentary democracy as the means of enacting legislation. Hayek defined democracy as a "means rather than an end", as democracy is "the only form of government that protects our individual liberty" (Hayek, 1960/2011), and "as the only effective method which we have yet discovered of making peaceful change possible" (Hayek, 1979). However, Hayek was concerned about the potential of unconstrained democracy to discredit democracy – for instance, under the influence of interest groups. For the same reason, neoliberals do not believe in democratic deliberation. They prefer to protect democratic rights by adopting constitutional constraints – for example, by restricting the power to legislate (Vallier, 2022).

As such, neoliberalism does not have a preferred theory of justice, as neoliberals appeal to a plurality of moral considerations and cultural and political economic practices to justify their preferred institutions (Vallier, 2022). All neoliberals believe in a person's right to own private property, including the right to own capital (Hayek, 1988). They also believe in the right of all persons to be treated as equals, but they all reject the idea of a welfare state, as they believe that a government should practice modest taxation and redistribution of wealth, and provide public goods, and social insurance for poor people. Hayek believed that a person's equality should be secured by a society's strong legal, administrative, and political institutions protecting the rule of law. However, Hayek rejected the idea of social justice involving a specific distribution of natural resources or precise income distribution by the state (Caldwell, 2004), as he was concerned that the pursuit of social justice through tinkering with particular economic outcomes may lead to compromising the rule of law (Hayek 1960/2011). Friedman also rejected the idea of "sharing the wealth" in order to reduce inequality, and believed that the way to reduce social inequality consisted in improving the workings of the market, by strengthening competition and widening opportunities for individuals to make the most of their own qualities (Friedman (1955). As for Buchanan, a contractarian economist, he did not believe in natural rights, as his idea of relative morality, borrowed from Hume, was that morality is "what we agree to" (Buchanan, 1975). Consequently, like Rawls, he defined "social justice as fair distribution of the benefits and burdens of social cooperation", provided that the principles of distribution were agreed upon by the whole society. John Rawls developed his model of social justice in the book *A Theory of Justice* published in 1971. Since then, his model has been accepted as the ideal theory of just distribution in the Global North high-income societies. How does his model account for the increased rates of poverty as well as income and wealth inequality in the very same societies? A 2014 OECD report stated that:

> The gap between rich and poor is now at its highest level in 30 years in most OECD countries. Today, the richest 10% of the population in the OECD area earn 9.5 times more than the poorest 10%. By contrast, in the 1980s the ratio stood at 7:1.
>
> (OECD, 2014)

In 2021, an Economic Policy Institute report noted that the pay of CEOs of the largest public firms in USA has skyrocketed 1,322 per cent since 1978 (Mishel and Kandra, 2021). There are several reasons why John Rawls' theory of justice fails to eliminate inequality and poverty in a liberal democracy. First, Rawls developed his theory of justice with reference to an ideal "well-ordered society", in which everyone accepts "the same principles of justice", and in which "the basic social institutions" generally satisfy these principles (Rawls, 1971). Such a "well-ordered" society does not exist in practice. In real life, some people are moral, some are immoral; some people comply with the rule of law, some do not; some people believe in a right to own property, some believe in a right to have their needs satisfied, and some others choose to act unjustly in terms of their interpersonal relations. Rawls himself acknowledged, in his 1993 book *Political Liberalism*, that in a modern liberal democracy, a "well-ordered society" "will be a society in which citizens embrace a diversity of reasonable comprehensive doctrines, religious and otherwise, many of which have their own distinctive way of thinking about justice" (Wolterstorff, 2013). Crafting just rules and institutions for such a society "which lacks genuine moral consensus" (MacIntyre, 1981) and whose members are not all ready to comply with the rules, is problematic, if not impossible. Second, the principles of justice in Rawls' ideal society were chosen by rational agents placed "behind a veil of ignorance" concerning their condition in life. The two principles of justice these agents agreed on are: (1) equal rights to basic liberties; and (2) the solution of social and economic inequalities will be crafted to the greatest benefit of the least-advantaged, "consistent with the joint savings principle", by "offices and parties open to all" and offering "fair equality of opportunity" (Rawls, 1971). As Rawls' model rejects a welfare state solving these social and economic inequalities, his fair distribution principle states:

> All social primary goods – liberty and opportunity, income and wealth, and the bases of self-respect – are to be distributed equally unless an unequal distribution of any or all of these goods is to the advantage of the least favoured.
>
> (Rawls, 1971)

Rawls did not explain why this theory of distribution is just; he simply assumed "that the long-grown moral tradition of our culture, based on our de facto intuitions, is right" (Ekardt, 2020). Rawls' "veil of ignorance" experiment is based on the questionable belief that mature, reasonable individuals do not possess "personal knowledge" (Polanyi, 1958) which will impact their choices. In addition, the way Rawls' theory of justice is currently applied in social policies narrows down the individual's freedom to economic equality with respect to individual needs to be satisfied by redistribution of income or wealth, on the simple assumption that the poor individual's needs are justified and that their deprivation is *undeserved*. However, as MacIntyre noted in 1981, "the notion of desert is at home only in the context of a community whose primary bond is a shared understanding both of the good for man and of the good of that community". Rawls' theory of justice assumes away the "ideal" society and solely focuses on individuals' self-interest which must be satisfied within a social contract based on an institutional arrangement. We can conclude that Rawls' theory of justice does not secure genuine *intragenerational* justice, his solution being just a technical means to keep a relative social peace in a socially disjointed neoliberal society. His theory of justice cannot be applied to *intergenerational* justice, either. As we have seen, Rawls included in his second principle the provision of "the joint savings principle", which refers to fair investment in the interest of future generations. However, in the same way that he did not explain the reason why his distributional model is just for the current generation, "the Rawlsian model provides no reason why justice should depend on basic orders opening up in terms of time" (Eckardt, 2020) over generations. While Rawls' theory of justice as "fairness" is a theory describing an ideal liberal democracy, whose "right order" is assumed, when we talk about sustainability we need to start from our current real, unsustainable world in which societies are built around unlimited-growth economies and hold a truncated relativistic morality that has produced many scared, unhappy individuals, leading precarious lives, due both to the current climate crisis and the social crises of inequality.

Looking beyond Rawls' theory of justice inadequacy, we need to ask a deeper question: How could the neoliberal ideas, initially formulated by eminent political economists with the goal to warn liberal societies against the dangers of fascist and socialist totalitarianism, transform the advanced capitalist societies in the last five decades, making them unsustainable both ecologically and socially? In 1944, Karl Polanyi's prescient explanation was that the belief in self-regulating markets has led to the sovereignty of the market over the non-market side of society, and the process of building a market society was encouraged by the governments of liberal democracies, as "the market has been the outcome of a conscious and often violent intervention on the part of government which imposed the market organization on society for noneconomic ends" (Polanyi, 1944). Polanyi's argumentation that in any society there is a social non-market side, in which the market system is contained, is well reasoned: the non-market side consists of the natural environment as well as of human beings engaged in non-contractual relations built within families, neighbourhoods, communities, and churches. These non-contractual relations secure social reproduction, socialization of children and the young, strong communities, and the creation and transmission of affective dispositions, and of shared meanings and values that underpin social cooperation (Fraser, 2014). The non-market side of society is essential for the existence of the market system of the economy to which it is supplying "their background conditions of possibility" (Fraser, 2014), such as natural resources, labour force, human and social capital, etc. As the unrestrained market system expanded, driven by the inherent propensity of capital to self-expand through economic growth and technological innovation, as well as through a lack of meaningful government control of markets and of powerful corporations and lobby groups, the market system started to encroach on the non-marketized side of society, first through an overuse of natural resources, followed by the capital domination of waged labour. Powerful corporations, enjoying personhood status, and freed from government control in the name of "deregulation", defined as "a process in which businesses would be disciplined by competitive market forces rather than public agencies" (Magnuson, 2008), have succeeded in towering over capitalist societies. Of the 30 million businesses operating in the USA, 5 million are corporations, yet they account for the majority of economic production, and 1 per cent of the corporations receive more than two-thirds of all corporate incomes (Magnuson, 2008). The rise of the financial sector which started to dominate the national economies and expanded globally between 1970 and 2008 has clearly shown the need for government protection of the non-market side of the economy, at a time when governments' priorities focused on helping markets function well. "Robbed of the protective covering of cultural institutions, human beings would perish from the effects of social exposure; they would die as the victims of acute social dislocation through vice, perversion, crime and starvation", warned Polanyi in 1944. The state protection of the non-marketized entities and institutions started to diminish in the 1970s free-market economies, with major cuts in social spending on public goods such as health care, education, arts, and culture, and an accelerated process of secularization, with public policies assaulting social reproduction (Fraser, 2014) and keeping faith and churches out of the public sphere. If left unchecked, the social and political interventions which have weakened the social fabric organically growing in the non-market side of any society will eventually lead to the demise of the whole society built on sovereignty of unrestrained markets. Globalization and the unlimited power yielded to global elites today is rapidly spreading and heightening these economic, ecological, and social risks to the whole global society. McCarraher vividly describes the dystopian global society dominated by the logic of

> Market Everlasting: unfettered free trade and globalization, epitomized by the formation of the World Trade Organization in 1995; the privatization of public services and their lucrative relegation to the caprices of the marketplace; the "reinvention" of government agencies along the line

of corporate bureaucracies; and the maniacal deregulation of finance and industry under the rubric of "modernization", liberating corporations from "stifling" restrictions on their freedom to invest, pollute and exploit. Under the aegis of the World Trade Organization, the World Bank, and the International Monetary Fund, globalization built a paradise of capital, with its incandescent minarets ascending on every continent, looming over multitudes pauperized into despair, impressed into the legions of wage servitude. Or sedated by the opiates of the spectacle.

(McCarraher, 2019)

Hayek, Friedman, and Buchanan should not be held accountable for the dystopian, unsustainable societies that characterize the advanced neoliberal democracies today. Blame should be laid on the politicized use of their ideas by political actors and political heads of governments who have abdicated their role as providers of public goods and have used their power to stifle public deliberation in national contexts. In our market societies, life is further complicated by the rapid advances and dissemination of information technologies, with potential to marginalize groups which do not have access to broadband transmission technologies. And since everything tends to be digital, governments have to build everything according to Silicon Valley IT standards – in the cloud – controlled only by a few big players like Amazon, Microsoft, and Google, and everyone else needs to play according to strict (and complex) digital standards that only a few know how to operate and use. The social cohesion has also unravelled under the impact of secularization, which has clearly divided society in a "post-virtue" (MacIntyre, 1981) public arena, where God is considered officially dead and replaced by various consumerist or ideologized idols, nothing is sacrosanct, and people are free to choose their morals, and a private arena, where traditional institutions are weakened by public policies crafted for the secularized public arena. The results are weak communities, poorly educated children, health care systems in crisis, broken families, and confused individuals perplexed by a range of undeserved inequalities. These inequalities have given rise to radical ideologies leading to social divisions and unfair questioning of the values of representative democracy and liberalism. For instance, those fighting for social justice expect society to solve a wide range of diverse issues such as racism, equity, education, healthcare, immigration, and LGBTQ+ rights in the name of securing equal treatment for everyone. While some of these battles are justified, some of them are impossible to win, as they do not have human solutions. It is impossible to expect that a woman competing in a sports competition against a male self-identifying as a female will have any chance of winning in a brute physical force contest. As Thomas Sowell states, some social justice advocates are seeking "cosmic justice" by attempting to correct the deficiencies not merely of society, but of the cosmos. They are trying "to correct not merely the sins of man but the oversights of God or the accidents of history" (Sowell, n.d.). Many of the social justice issues are produced by individuals making choices outside of the universally accepted patterns of existence and expecting society to solve these social issues by accepting their vision of equality.

The existing global solutions proposed for saving the planet from climate change, namely *green growth*, the focus of the 2015 Paris Agreement aiming to mitigate national carbon emissions by 2050, and *inclusive growth*, the centrepiece of the UN global Sustainable Development Goals, do not come close to solving the problem of unsustainability, as they do not deal with capitalism's inherent need for economic growth and its economic, social, and ecological consequences. These solutions found to date, while citing intergenerational justice as a normative goal, do not provide a valid answer to the problem of intragenerational justice, manifested in rampant inequality between the high-income countries of the Global North and low- and medium-income countries of the global South, and within most countries in the world, producing poverty, unrest, and despair among real human beings. The new report to the Club of Rome entitled *Earth for All: A Survival Guide for Humanity* (Dixson-Declève et al., 2022) uses a system dynamics model called "Earth4All" to explore the potential of a scenario called the Giant Leap to transform the global economic system by 2050 in a way that will "push the global economy off the destructive course it is on and onto a resilient path" with "a stable global economic system that delivers higher wellbeing to the majority". Among the "five extraordinary turnarounds" proposed for securing "global equity on a healthy planet" (ending poverty, addressing gross inequality, empowering women, making our food system healthy for people and ecosystems, and transitioning to clean energy), drastic interventions are proposed for income and wealth redistribution. These interventions include policies to limit to 40 per cent the amount that the wealthiest 10 per cent can take of national incomes, higher taxes on inheritances, and creation of Citizens' Funds made up of fees charged to the private sector "for extracting and using resources that should be seen as under the stewardship of all in society, including fossil fuels, land, freshwater, the ocean, minerals, the atmosphere, and even data and knowledge" (Dixson-Declève et al., 2022). The Citizens Funds' revenues will be equally redistributed to all citizens in a country through a "universal basic dividend (UBD)" (Dixson-Declève et al., 2022). While the measures proposed in the Earth4All initiative may work to lower economic inequality, they are not the optimal solution for solving the global justice unsustainability issue, as they do not address in any way the moral dilemmas of sustainability. We propose a theory of justice centred on human beings' right to exist as human beings and not as *Homo economicus* or in human capital form only, to replace the mechanistic instrument of "just distribution" imagined by John Rawls in 1971, and whose echoes are still found in the Earth4All initiative. To find a more encompassing definition of justice, going beyond distributive justice, we will turn to the ancient Greek philosophers' concept of justice.

From right-order to rights-based justice

Ancient Greek philosophers defined political justice as a *social virtue* essential for the peace and order in an ideal democracy governed by the rule of law. For Plato and Aristotle, justice "is found among men who share their life with a view to self-sufficiency, men who are free and either proportionately or arithmetically equal" and "whose mutual relations are governed by law" (Aristotle, 1947). In his *Republic*, Plato (427–347 BC) defined justice as the greatest relational virtue found "in the dealings of citizens with one another" (Plato, 1947). He also defined distributive justice, a type of justice referring to the allocation of rights, duties, and responsibilities among the members of a community. Noting that "democracy distributes an odd sort of equality to equals and unequals", Plato believed that unequal distribution of resources is indispensable to equalize benefits in the case of unequals. He called this type of distribution *just distribution*, not equal distribution of benefits, and mentioned that in a democracy the governments should secure just distribution (Plato, 1947). Plato's student Aristotle (384–322 BC), in his *Nicomachean Ethics*, added that just distribution should be "according to merit", but he acknowledged that "merit" will be defined according to a person's contribution to achieving "the good" in a community, where "the good" can be either freedom, wealth, or excellence of character, depending on the political orientation of the community: "democrats identify it with the status of freeman, supporters of oligarchy with wealth (or with noble birth), and supporters of aristocracy with excellence" (Aristotle, 1947). Building on Plato's idea that justice is "another's good" which all others are bound to observe, Roman lawyer Ulpian (AD 211–222) defined justice as "the steady and lasting willingness to give to others what they are entitled to" (their *ius*, or right): "*constans et perpetua voluntas ius suum cuique tribuere*" (Justinian, 1896). In Roman law, justice was one of the three precepts consecrated in the legal code compiled by the Byzantine Emperor Justinian I: "to live honestly, to injure no one, and to give every man his due" (Justinian, 1896). Ulpian's definition of justice, implying *a duty* to everyone without discrimination, as well as the existence of correlative *rights* that belong to everyone indifferently, has introduced a rights-based justice in the philosophical discourse as well as in the legal doctrine and subsequent practice. This rights-based theory of justice was perfected in the thirteenth century by the medieval philosopher-theologian Thomas Aquinas (1224/5–1274). Aquinas started from the idea that each person is inherently endowed with the "natural light of reason" and with moral impulses which enable them to know the primary truth that "good is to be done and pursued, and evil is to be avoided". He defined "the good" to be pursued, first as natural goods that make life possible, starting with "life, the procreation and education of offspring, knowledge and a civil social order." As for justice, Aquinas defined it as an individual virtue of character which can be achieved by perfecting one's will, as justice is a "habit according to which one gives to everyone what is right (*ius*) with a constant and perpetual will" (Aquinas, 2012). In Aquinas' view, an individual's virtue of justice, when perfected, contributes to justice in society by aligning the forms and norms of legal justice with the general good (Dierksmeier, 2011). How can human beings perfect their justice virtue? Thomas Aquinas accepted Aristotle's metaphysical teleology by believing that all natural beings are inclined, through their natural propensities, to achieve "their respective, proper good". As for humans, he believed that all human beings are naturally equipped to pursue the natural goods of human life, and some will even consider the "quest for the perfection of the soul" (*beatitudo*) (Dierksmeier, 2011). Through his moral philosophy, based on natural law, Aquinas shed light on the potential of all human beings to conceive and pursue higher goals than the contemporary – much narrower – self-interest illustrated by the "preferences" of *Homo Economicus*, as revealed by markets. We can use Aquinas' insights to develop a rights-based theory of justice as the centrepiece of a narrative for a sustainability-oriented society in which individuals are able to live according to life-enhancing moral norms and policies, and using the market simply for its initial purpose: the efficient allocation of resources. Accepting a rights-based sustainability theory of justice requires an open mind and recognition that the modern theory of justice developed by the Enlightenment philosophers paints an incomplete and frankly false picture of human beings in their social interactions. For instance, the Scottish philosopher David Hume (1711–1776), defined justice as an "artificial" virtue, as for him to be "just" does not occur naturally in humans, and "the sense of justice and injustice is not deriv'd from nature, but arises artificially [...] from education, and human conventions" (Hume, 1751). For Hume, a system of justice is essentially "a relatively settled set of conventional expectations between individuals chiefly concerned with their own interest that is reinforced by sentiments of approval and disapproval". There are no absolute principles of right and wrong in Hume's theory of justice, as for him "human behaviour is guided by human feelings and preferences which are considered 'right' *per se*" (Ekardt, 2020). Likewise, Adam Smith considered justice as a "negative virtue" which "often all it requires is that we sit still and do nothing" (Smith, 1759/1984). This means that a person can theoretically "sit still and do nothing" even as they see a human being attacked and robbed in the street. Adam Smith's theory of justice is still based on human preferences, even as Smith himself stated that human interactions are motivated not only by self-interest but by "sympathy" and "fellow-feeling" sentiments (Smith, 1759/1984). If we accept that justice and the rule of law are essential to the sustainability of a democratic society, we must accept Thomas Aquinas' theory of justice based in the natural law, as the natural law "does not represent some naïve naturalism" (Dierksmeier, 2011), but is rooted in the objective order that exists in nature and which gives human beings some universal and inalienable natural rights, such as the right to "life, liberty and the pursuit of happiness", as stated in the United States' Declaration of Independence. These natural rights cannot be repealed by human laws or conventions.

Sustainability beyond distributive justice

In 2020, Felix Ekardt wrote: "The world is physically finite, and humans are biological beings that cannot exist without intact ecosystems, fertile soils, drinkable water and a reasonably intact global climate" (Eckardt, 2020). He also added that "The political, legal and moral order of Western states has enabled most citizens living there to enjoy a historically unique degree of freedom and prosperity" (Ekardt, 2020).

However, this order has provided a "primarily economy-oriented under-standing of freedom" (Ekardt, 2020) based on "needs satisfaction" (Gough, 2017). Ekardt proposed a new ethical and legal theory of sustainability based on a wider interpretation of human rights, including the right to elementary preconditions of freedom, such as "life, health, subsistence level in the form of food, water, security, climate stability, elementary education, absence of war, and civil war, etc." (Ekardt, 2020). Ekardt notes that these rights to freedom should be protected in all liberal democratic states, as well as in supranational entities, such as the European Union, and in all international organizations and institutions, such as the United Nations Organization, by using a legal and ethical balancing theory based on liberal democratic principles. Ekardt also referred to frugality as a necessary (but not the only) sustainability strategy, given the considerable issues of a post-growth world (Ekardt, 2020). Having personally experienced Ceausescu's totalitarian regime in communist Romania before 1989, I agree with Ekardt that a functioning liberal democ-racy, with a responsible government controlling the excesses of markets and corporations, and protecting the human right to life and its preconditions, is the best setting for implementing sustainability. I also agree that human beings are complex creatures and that it is wrong to treat them simply as producers and consumers of goods and services aiming to maximize profits and utility in free-market economies. Human beings have more than eco-nomic rights; they have a right to life; the right to be free from interventions that tend to destroy life and the decent conditions of life; they have a right to think for themselves and to express their thoughts and beliefs; a right to work and be honestly valued for their contribution; and a right to be treated as dignified social beings. To put it simply, human beings have the right to live and act in order to fulfil their two "most important purposes: to stay alive and keep a place under the social sun" (Georgescu-Roegen, 1971). These universal human rights have been established by the United Nations Organi-zation after humanity experienced the horrors of the Second World War. The United Nations International Bill of Human Rights (the Universal Declara-tion of Human Rights) was established in 1948, and it was followed by the International Covenant on Civil and Political Rights and the International Covenant on Economic, Social and Cultural Rights, both established in 1966. While the United Nations' International Bill of Human Rights does not define human rights, it proclaims in Article 1 that "all human beings are born free and equal in dignity and rights. They are endowed with reason and conscience and should act towards one another in a spirit of brotherhood." (Universal Declaration, 1948). The document includes as human rights the right to life and liberty, to security of persons with freedom from slavery and torture, freedom of opinion and expression, the right to work, to get an education and to found a family, which is defined as "the natural and fundamental group unit of society and is entitled to protection by society and the State" (Universal Declaration, 1948). These rights are inherent to all human beings regardless of race, sex, nationality, ethnicity, language, religion, or any other status. Everyone is entitled to these rights, without discrimination. These United Nations *foundational* documents imply that a human right is a "right such that the status sufficient for possessing the right is that of being a human being" (Wolterstorff, 2013). One does not need to be a virtuous human being to be granted human rights. While these rights are grounded in "inherent human dignity", the United Nations' documents do not explain what accounts for human dignity. I propose that human dignity is grounded in human worth or the value of human beings, as worthy participants in the web of life objectively existing on the Earth. This participation is subject to some objective natural rules and laws, such as the laws of thermodynamics, the law of gravity, or the law of protein folding in molecular biology (Dill and MacCallum, 2012), which no human being can contradict or control. Why, then, do some humans believe that they "are free to choose" the moral rules of their interaction with other human beings, or other living creatures, without damaging the universal life-support system? Are these "others" worthless? This is where the rules of rights-based justice become important.

> The other comes into my presence as a creature of worth; I likewise come into her presence as a creature of worth. On account of her worth, she has legitimate claims on me as how to treat her; on account of my worth, I have legitimate claims on her as to how she treats me.
>
> (Wolterstorff, 2013)

This truth acts as a first principle and forces humans to behave in a certain way in relation to other human beings, because, as Aquinas teaches, "good is to be done and evil is to be avoided" (Kaczor and Sherman, 2008). This is the core tenet of *primary justice*. Primary justice is a justice system that protects primary human rights defined as "rights that anyone with a rational plan of life would want for herself to pursue her conception of the good and justice, and ones she is willing to extend to others on reciprocal terms" (Vallier, 2019). For example, a primary right might be any person's right to enter a public health care institution and seek help during opening hours without being denied access. In his book *Journey Toward Justice*, Nicholas P. Wolterstorff (2013) also defined *reactive justice*. Reactive justice is a system of justice based on reactive rights, defined as "the complex of permission-rights and claim-rights that are acquired by virtue of someone wronging someone" or "rights acquired in reaction to being wronged" (Wolterstorff, 2013). Both primary and reactive justice are important for implementing sustainability at the level of society, where right actions and observance of the rule of law will lead to social peace and harmony, and in interpersonal relations, as "primary justice is present in society insofar as the members of society stand to one another in the normative social relationship of being treated as they have a right to be treated" (Wolterstorff, 2013). It follows that justice is bestowed on us not by an abstract objective standard defined by human beings but because of who we are, living creatures endowed with reason, conscience, and moral impulses, who know what is right and what is wrong, and have the right to exist, even if we do not always choose to do what is right. This is where the reactive justice is called in, to protect individuals and society against wrong-doing and to right the wrongs done. This understanding of sustainability as human-rights-based justice must be rooted in a moral and legal system that protects the right to a sustainable life of everyone equally. The way the moral and the legal systems of late capitalist societies has evolved disintegrates the existence of limits, both physical and moral. These boundaries do exist, however, and they were likely set up to keep human beings from self-destruction. A sustainable world is built starting from a lucid examination of and an acceptance of these boundaries, as well as from the development of a new narrative about physical and moral sustainability. This new narrative includes the facts of thermodynamics – indicating that a sustainable economy is "an economy with a constant stock of capital, capable of being maintained by a rate of material and energy throughput that is within the regenerative and assimilative capacities of the ecosystem." (Jackson, 2018). It also includes facts of the "whole human being" made as both wonderful and miserable beings, that are in need of objective moral values to guide their human goals and aspirations, and knowing that "the ultimate goal itself and the longing to reach it must come from another source" (Einstein, 1954). Only once a new sustainability narrative is formulated and widely accepted can we hope that new sustainability practices will emerge with the ability to change, for instance, the way we make investments and develop social policies. These practices should promote sustainability both by enhancing human dignity through socially meaningful projects and by protecting the environment from excessive consumption of the Earth's natural resources. If humans have an elementary right to education and work, more investment in every society should be directed to enhancing human knowledge and innovation, by producing more ideas, not only physical capital, according to Solow's growth theory (1956). One such investment is suggested by the endogenous growth theories (Romer, 1990; Aghion and Howitt, 1992; Lucas, 2002). These theories explain how economic growth through indefinite investment in education and research has spillover effects on the economy and is potentially able to reduce the diminishing marginal returns to capital accumulation, inherent in Solow's exogenous growth theory. The endogenous growth theories are macroeconomics models with two sectors: (1) a sector of *producers of final output* and (2) *a research and development sector* which develops ideas and technological innovation. One popular endogenous growth model is Arrow's "learning by doing" model (Arrow, 1962) based on incorporating in new goods all the knowledge available which has been obtained through accumulated experience. Another example is Edmund Phelps' "grassroots innovation" model (Phelps, 2013) of the 1850–1950 era, which in his view was the driver of a "dynamic economy", where "mass" innovation "bubbled up from countless resourceful and creative entrepreneurs". In the right cultural environment, these entrepreneurs could unleash "a technological tidal wave that could upset the status quo, kick out entrenched interests, and make daily life more interesting, more fluid, more laden with discovery and fulfilling and challenging work" (Mokyr, 2014) – an age of "mass flourishing" (Phelps, 2013). More investment not only in technological innovation but also in social sciences, in art, sport and in cultural projects could achieve J.S. Mill's dream of improving humans' "Art of Living" in harmony with nature and with each other.

This chapter has discussed the justice aspects of sustainability. Noting that currently the sustainable development concept only refers to intergenerational justice as a normative requirement, and that intragenerational justice normally administered by liberal democracies using John Rawls' theory of "fair" distributive justice is inadequate to "right" social inequalities and the overuse of natural resources, a new understanding of sustainability as justice is proposed. Starting from the idea that a basic principle of justice is embedded in the natural sustainability order existing on planet Earth, which protects all life equally, the chapter has discussed the concept of sustainability as primary and reactive justice rooted in the human right to life. Implementation of just sustainability requires a change of the current development narrative and subsequent practices, by taking into account the limiting physical and moral boundaries that underlie life on Earth.

References

Aghion, P. and Howitt, P. (1992) "A Model of Growth Through Creative Destruction." *Econometrica*, 60: 323–351.

Aquinas, T. (2012) *Summa Theologica*. Rockford, IL: Emmaus Academic.

Aristotle (1947) "Nicomachean Ethics." In *Man and Man: The Social Philosophers*, edited by S. Commins and R.N. Linscott. New York: Random House.

Arrow, K.J. (1962) "The Economic Implications of Learning by Doing." *Review of Economic Studies*, 29: 153–173.

Bainbridge, S.M. (2023) *The Profit Motive. Defending Shareholder Value Maximization*. Cambridge: Cambridge University Press.

Beckerman, W. (1998) "Sustainable Development: Is It a Useful Concept?" In *The Environmental Ethics and Policy Book: Philosophy, Ecology, Economics*, 2nd ed., edited by D. Van DeVeer and C. Pierce. Belmont, CA: Wadsworth Publishing Company.

Biermann, F., Hickmann, T., and Sénit, C. (eds) (2022) *The Political Impact of the Sustainable Development Goals: Transforming Governance Through Global Goals?* Cambridge: Cambridge University Press.

Buchanan, J. (1975) *The Limits of Liberty: Between Anarchy and Leviathan*. Chicago: University of Chicago Press.

Buchanan, J.M. (1979) *What Should Economists Do?* Indianapolis: Liberty Fund.

Burke, E. (1910) *Reflections on the Revolution in France*. London: Dent.

Business Roundtable (2019) "Business Roundtable redefines the purpose of a corporation to promote 'an economy that serves all Americans'." www.businessroundta ble.org/business-roundtable-redefines-the-purpose-of-a-corporation-to-promote-a n-economy-that-serves-all-americans.

Caldwell, B. (2004) *Hayek's Challenge*. Chicago: University of Chicago Press.

Chernilo, D. (2017) "The question of the human in the Anthropocene debate." *European Journal of Social Theory*, 20(1): 44–60.

Czech, B. (2002) *Shoveling Fuel for a Runaway Train: Errant Economists, Shameful Spenders, and a Plan to Stop Them All*. Berkeley, CA: University of California Press.

Daly, H. (1973) *Toward a Steady State Economy*. San Francisco, CA: W.H. Freeman.

Daly, H.E. (1998) "On Wilfred Beckerman's Critique of Sustainable Development." In *The Environmental Ethics and Policy Book: Philosophy, Ecology, Economics*, 2nd ed., edited by D. Van DeVeer and C. Pierce. Belmont, CA: Wadsworth Publishing Company.

Davis, J., Ossowski, R., Richardson, T., and Barnett, S. (2000) "Fiscal and Macroeconomic Impact of Privatization." Occasional Paper No. 194. Washington: International Monetary Fund.

Dierksmeier, C. (2011) "Thomas Aquinas on Justice as a Global Virtue." http://dx.doi.org/10.2139/ssrn.1737561.

Dill, K.A. and MacCallum, J.L. (2012) "The Protein-Folding Problem, 50 Years On." *Science*, 338: 1042–1046.

Dixson-Declève, S., Gaffney, O., Ghosh, J., Randers, J., Rockström, J., Stoknes, P.E. (2022) *Earth for All. A Survival Guide for Humanity. A Report to the Club of Rome Fifty Years After The Limits to Growth*. Gabriola Island, BC: New Society Publishers.

Einstein, A. (1954) *Ideas and Opinions*. New York: Crown.

Ekardt, F. (2020) *Sustainability Transformation, Governance, Ethics, Law*. Cham, Switzerland: Springer.

Esping-Andersen, G. (1990) *The Three Worlds of Welfare Capitalism*. Princeton, NJ: Princeton University Press.

Fraser, N. (2014) "Behind Marx's hidden abode." *New Left Review*, 86: 55–72.

Friedman, M. (1955) "Liberalism, Old Style." In *1955 Collier's Year Book*. New York: P.F. Collier & Son, pp. 360–363.

Friedman, M. (2002) *Capitalism and Freedom*. Chicago: University of Chicago Press. (Original work published 1962)

Gale, F.P. (2018) *The Political Economy of Sustainability*. Cheltenham, UK and Northampton, MA: Edward Elgar.

Georgescu-Roegen, N. (1971) *The Entropy Law and the Economic Process*. Cambridge, MA: Harvard University Press.

Goodland, R. (1995) "The concept of environmental sustainability." *Annual Review of Ecology and Systematics*, 26: 1–24.

Gough, I. (2017) *Heat, Greed and Human Need Climate Change, Capitalism and Sustainable Wellbeing*. Cheltenham, UK: Edward Elgar.

Hayek, F.A. (2007) "The Road to Serfdom." In *The Collected Works of F.A. Hayek*, Vol. 2, edited by B. Caldwell. Chicago: University of Chicago Press. (Original work published 1944)

Hayek, F.A. (2011) "The Constitution of Liberty." In *The Collected Works of F.A. Hayek*, Vol. 17, edited by R. Hamowy. Chicago: University of Chicago Press. (Original work published 1960)

Hayek, F.A. (1978) *Law Legislation and Liberty Volume 2: The Mirage of Social Justice*. Chicago: University of Chicago Press.

Hayek, F.A. (1979) *Law Legislation and Liberty Volume 3: The Political Order of a Free People*. Chicago: University of Chicago Press.

Hayek, F.A. (1988) *The Fatal Conceit: The Errors of Socialism*, edited by W.W. Bartley. Chicago: University of Chicago Press.

Hicks, J.R. (1950) *Value and Capital*. Oxford: Oxford University Press.

Hueting, R. and Reijnders, L. (1998) "Sustainability as an Objective Concept." *Ecological Economics*, 27(2): 139–147.

Hume, D. (1975) "Enquiry Concerning the Principles of Morals." In *Enquiries*, edited by P.H. Nidditch. Oxford: Clarendon Press. (Original work published 1751)

ILO-UNICEF Office of Research (2022) *The role of social protection in the elimination of child labour: Evidence review and policy implications*. Geneva and Florence: International Labour Organization and UNICEF Office of Research/Innocenti.

Jackson, T. (2018) "How the light gets in." November 12, 2018. www.resilience.org/stories/2018-11-12/how-the-light-gets-in.

Justinian (1896) *The Institutes of Justinian*, translated by J.B. Moyle. Oxford: Clarendon Press.

Kaczor, C. and Sherman, T. (2008) *Thomas Aquinas on the Cardinal Virtues: A Summa of the Summa on Justice, Courage, Temperance, and Practical Wisdom*. Washington, DC: Catholic University of America Press. https://doi.org/10.2307/j.ctv194cpdf.

Kaplan, S. (2019) *The 360° Corporation: From Stakeholder Trade-Offs to Transformation*. Redwood City, CA: Stanford Business Books.

Kelton, S. (2020) *The Deficit Myth: Modern Monetary Theory and the Birth of the People's Economy*. New York: PublicAffairs Books.

Keynes, J.M. (1965) *The General Theory of Employment, Interest, and Money*. New York: Harcourt, Brace. (Original work published 1936)

Lucas, R.E. (2002) *The Industrial Revolution: Past and Future, Lectures on Economic Growth*. Cambridge, MA: Harvard University Press.

MacIntyre, A.C. (1981) *After Virtue: A Study in Moral Theory*. Notre Dame, IN: University of Notre Dame Press.

Magnuson, J. (2008) *Mindful Economics: How the US Economy Works, Why It Matters, and How It Could Be Different*. New York: Seven Stories Press.

McCarraher, E. (2019) *The Enchantments of Mammon: How Capitalism Became the Religion of Modernity*. Cambridge, MA: Belknap Press of Harvard University Press.

McNeill, J.R. and Engelke, P. (2014) *The Great Acceleration: An Environmental History of the Anthropocene Since 1945*. Cambridge, MA: Belknap Press of Harvard University Press.

Metcalf, S. (2017) "Neoliberalism: the idea that swallowed the world." *The Guardian*, August 18, 2017. www.theguardian.com/news/2017/aug/18/neoliberalism-the-idea-that-changed-the-world.

Mill, J.S. (1986) "John Stuart Mill on the Stationary State (1848)." *Population and Development Review*, 12(2): 317–322.

Mirowski, P. (2014) "The Political Movement that Dared not Speak Its Own Name: The Neoliberal Thought Collective Under Erasure." Working Paper, No. 23. New York: Institute for New Economic Thinking.

Mishel, L. and Kandra, J. (2021) "CEO pay has skyrocketed 1,322% since 1978." Economic Policy Institute. www.epi.org/publication/ceo-pay-in-2020.

Mokyr, J. (2014) "A Flourishing Economist: A Review Essay on Edmund Phelps's Mass Flourishing: How Grassroots Innovation Created Jobs, Challenge, and Change." *Journal of Economic Literature*, 52(1): 189–196.

Mosler, W. (2013) *Soft Currency Economics II: The Origin of Modern Monetary Theory.* CreateSpace Independent Publishing Platform.

Niebuhr, R. (1960) *Moral Man and Immoral Society A Study in Ethics and Politics.* New York: Charles Scribner's Sons. (Original work published 1932)

OECD (2011) *Divided We Stand: Why Inequality Keeps Rising.* Paris: OECD. www.oecd.org/els/soc/dividedwestandwhyinequalitykeepsrising.htm.

OECD (2014) "Focus on Inequality and Growth – December 2014." www.oecd.org/social/Focus-Inequality-and-Growth-2014.pdf.

Phelps, E. (2013) *Mass Flourishing: How Grassroots Innovation Created Jobs, Challenge, and Change.* Princeton, NJ: Princeton University Press.

Piketty, T. (2014) *Capital in the Twenty-First Century,* Cambridge, MA: Harvard University Press.

Plato (1947) "The Republic." In *Man and Man: The Social Philosophers,* edited by S. Commins and R.N. Linscott. New York: Random House.

Polanyi, K. (1944) *The Great Transformation.* Boston, MA: Beacon Press.

Polanyi, M. (1958) *Personal Knowledge: Towards a Post-Critical Philosophy.* Chicago: University of Chicago Press.

Rawls, J.B. (1971) *A Theory of Justice.* Cambridge, MA: Harvard University Press.

Rawls, J.B. (1993) *Political Liberalism.* New York: Columbia University Press.

Romer, P.M. (1990) "Endogenous Technological Change." *Journal of Political Economy*, 98: S71–S102.

Schumpeter, J.A. (1934) *Theory of Economic Development.* Cambridge, MA: Harvard University Press.

Smith, A. (1984) *The Theory of Moral Sentiments.* Indianapolis: Liberty Fund. (Original work published 1759)

Solow, R. (1956) "A Contribution to the Theory of Economic Growth." *Quarterly Journal of Economics*, 70(1): 65–94.

Solow, R. (1991) "Sustainability: An Economist's Perspective". In *The Environmental Ethics and Policy Book. Philosophy, Ecology, Economics* (2nd edition), edited by D. VanDeVeer and C. Pierce. Belmont, CA: Wadsworth Publishing Company.

Sowell, T. (n.d.) "The Quest for Cosmic Justice." www.tsowell.com/spquestc.html.

Steffen, W., Broadgate, W., Deutch, L., Gaffney, O., and Ludwig, C. (2015) "The trajectory of the Anthropocene: The Great Acceleration." *The Anthropocene Review*, 2 (1): 81–98.

Stout, L. (2012) *The Shareholder Value Myth: How Putting Shareholders First Harms Investors, Corporations, and the Public.* Berrett-Koehler Publishers.

Swan, T.W. (1956). "Economic growth and capital accumulation." *Economic Record,* 32(2): 334–361.

Universal Declaration (1948) The International Bill of Human Rights. Adopted and proclaimed by General Assembly resolution 217 A (III) of 10 December 1948. www.ohchr.org/sites/default/files/Documents/Publications/Compilation1.1en.pdf.

UNGA (2015) A/RES/70/1 "Transforming our world: the 2030 Agenda for Sustainable Development" Resolution adopted by the General Assembly on 25 September

2015. New York: United Nations Organization. www.un.org/en/development/desa/p opulation/migration/generalassembly/docs/globalcompact/A_RES_70_1_E.pdf.

Vallier, K. (2019) "Primary Rights." In *Must Politics Be War? Restoring Our Trust in the Open Society*. New York: Oxford Academic, pp. 156–172. https://doi.org/10.1093/oso/9780190632830.001.0001.

Vallier, K. (2022) "Neoliberalism." *The Stanford Encyclopedia of Philosophy* (Winter 2022 Edition), edited by E.N. Zalta and U. Nodelman. https://plato.stanford.edu/a rchives/win2022/entries/neoliberalism.

WCED (1987) *Our Common Future*. The United Nations World Commission for Environment and Development. Oxford: Oxford University Press.

Wolterstorff, N.P. (2013) *Journey Toward Justice Personal Encounters in the Global South*, Grand Rapids, MI: Baker Academic.

Wray, L.R. (1998) "Modern Money." Levy Economics Institute Working Paper No. 252, http://dx.doi.org/10.2139/ssrn.137409.

7 Sustainability as freedom

From a human perspective, sustainability means acceptance of the *real* natural ecological and social order which accounts for the continuation of life on Earth, and the adjustment of humans' lives and their socio-economic decisions to fit within the natural order without impairing this "gift of a working life-support system" (Murphy et al., 2021) and keeping it available for next generations. As the life-support system exists objectively, the first thing human beings need to do is to accept this gift with gratitude, while freely enjoying, for instance, humble perspectives offered by the sight of a wild dandelion, in the meek awareness that they could not themselves "have invented either the dandelion or the eyesight" (Chesterton, 1969). A second related notion is to carefully use this free gift, mindful of the complex and dynamic rules involved in its existence and functioning. The discovery of these rules and the limits they impose is a legitimate expectation from scientists, who are called both "to establish an appropriate maximum environmental burden" which will keep intact "the function of the environment and natural resources" (Hueting and Reijnders, 1998), and to indicate the limits to humanity's freedom needed to enjoy the life-sustaining system. This latter task is very difficult, given human nature and subjectivity in assessing what "enjoyment" and the "good life" means. Starting in the 1970s, natural scientists have established that there are objective natural limits, and have warned about the catastrophic consequences of unlimited economic growth on a non-expanding planet (Meadows et al., 1972; Wackernagel and Rees, 1996; Meadows et al., 2004). In the new millennium, scientists have indicated that some planetary boundaries have already been exceeded (Rockström et al., 2009; Steffen et al., 2015; Raworth, 2017). However, humans continue to ignore these warnings, and keep organizing their economic activity according to human-made expectations and rules, "largely uncoordinated, sometimes incompatible, and absent explicit reference to principles commensurate with a finite planet. Rather, decisions are dominated by short-term financial considerations that do not reflect biophysical realities or prioritize long-term sustainability" (Murphy et al., 2021), as our current economic and social order "is built to maximize the extraction of wealth and profit – from both humans and the natural world – a reality that has brought us to what we

DOI: 10.4324/9781003307587-8

might think of it as capitalism's techno-necro stage" (Klein, 2023). Failure to prioritize long-term sustainability has led to the current web of ecological, social, and humanitarian crises which threaten the very continuation of life on Earth. The real problem is that while establishing physical ecological limits to humans' economic activity is a necessary condition for sustainability, it is not a sufficient one, given that sustainability is not *a state*, but a continuously evolving *process* driven by the dynamism and intricacies of the socio-ecological interactions involved in the life-sustaining system. In assessing these intricacies, human behaviour inclined toward sustainability does count, because

> sustainability is ultimately an issue [not just of science but] of human behavior, and negotiation over preferred futures, under conditions of deep contingency and uncertainty. It is an inherently normative concept, rooted in real world problems and very different sets of values and moral judgements.
>
> (Robinson, 2004)

This fact raises some unavoidable questions about human freedom: first of all, what is it; and how free are humans to exceed the planetary boundaries without limiting their freedom to exist? What type of human freedom exists in the liberal democratic societies currently dominated by neoliberal thinking and rapidly evolving information technologies (IT)? What is the meaning of human freedom in a sustainable society? This chapter will answer these questions starting from two premises. The first one is that the longing for freedom, or for "pursuing our own good in our own way" (Mill, 1859), is one of the most powerful, universal drivers of human action. The truthfulness of this premise has aptly been articulated in a recent report which states that "More than anything else, five decades of Freedom in the World reports demonstrate that the demand for freedom is universal" (FIW, 2023). The second premise is that humans are social beings, and their freedom will always be impacted both by their biological features and by the interaction rules established in their society where "liberty is only that which legitimate authority recognizes" (Huffman, 1977). Still, free humans need to decide what choices they make. In 1762, Jean-Jacques Rousseau defined moral freedom as an individual's freedom to be "master of himself" by living according to rules rationally prescribed for oneself: "For the impulse of appetite alone is slavery, and obedience to the law one has prescribed for oneself is freedom" (Rousseau, 2012). Unlike Jean-Jacques Rousseau, I argue that individual freedom, particularly in a society aiming to be sustainable within this and future generations, involves a conscious commitment of each of its members to a way of living which clearly distinguishes between what a person "most wishes as an individual, and what he 'ought' to wish – between 'maximum satisfaction' in the economic jargon, and 'right' answers to questions" (Knight, 1941). These questions are vital questions about our condition of living beings inhabiting a life-friendly planet. A sustainable person will choose not only what they prefer to do, as utilitarianism prescribes, but also what they ought to do. A first condition for this to happen is an honest search for the truth, not only the evidence-based truth given by science, but also the truth from other sources (philosophy, ethics, theology) that can offer answers to fundamental questions about who we are as humans, why we are here, or what might happen when we die. A second condition would be to make the right choice and cultivate the virtues of a resilient self. These virtues are quintessential to living and flourishing in a free society – self-reliance, personal responsibility, respect for individual choice, and neighbourliness, "in a framework where religious observance helps nurture these virtues" (McGinnis, 2023) and just institutions protect them. Genuinely free individuals will be able to make sustainable choices which take into account not only their interests but also the interests of other living creatures. This chapter will begin with a discussion of human freedom, then consider freedom in liberal democratic nation states, whose core premises (human dignity, impartiality, and freedom protected by the institutionally balanced rule of law, and a democracy with parliaments and separation of powers), were at least theoretically designed to secure the freedom of all its members. Finally, I will discuss what freedom entails for a sustainable individual.

What is freedom?

Like any "essentially contested concept" (Gallie, 1955), human freedom is difficult to define. The first difficulty arises from the long-standing debate among philosophers and theologians whether freedom of the will exists or not. This debate has acquired paradoxical undertones since the 1920s when logical positivism and empiricism started to dominate in modern science, due to the "conventional fear of the transcendent or transcendental" (Sellars, 1946), aspects which were considered meaningless in scientific research. This has relegated freedom to the idealistic class of concepts which either do not exist or are just illusions. On one hand, there are hard-line materialists and determinists in science and behavioural social sciences who believe that "freedom of the will is an illusion" (Wilson, 1998) and that humans are not free to make decisions but "are somehow simply captives of tiny, self-copying entities called genes" (Ehrlich, 2000). On the other hand, there are those who believe that freedom is a value impacting human behaviour and social organization. This view has produced a body of literature that accepts that humans have free will and consequently are morally responsible beings (Kane, 1996). This literature also includes works by ecological economists who insist, for instance, that policy decisions about means and ends rule out determinism and nihilism, as humans making decisions are free to choose among alternatives and "that there is a real criterion of value to guide our choices" (Daly and Farley, 2011). It also includes evolutionary ecologists, such as Paul Ehrlich (2000).

In 1969, the philosopher Isaiah Berlin counted more than 200 definitions of freedom as recorded by historians of ideas (Berlin, 1969). They range from the idea that a free person enjoys freedom in all their choices – "Liberty is to live upon one's own terms; slavery is to live at the mere mercy of another" (Trenchard and Gordon, 1971) – to the idea that some choices must be given up in order to accommodate existing rules of social organization, or to prioritize other goals considered more valuable. In Berlin's words, freedom must be "compatible with the existence of organized society" (Berlin, 1969). A simple definition of freedom is "a right to act in the way you think you should" (Cambridge, 2020), given your desire to act in a certain way and the absence of any obstacle or restriction frustrating that desire. This type of *freedom from* interference by other human beings or by public authority is defined by Berlin as *negative freedom*. He considered negative freedom intrinsically valuable for not "degrading or denying our nature", given that by preserving "a minimum area of personal freedom" an individual is able to develop "his natural faculties which alone makes it possible to pursue, and even to conceive, the various ends which men hold good or right or sacred." (Berlin, 1969). Negative freedom in the form of basic entitlements to "freedom from fear and want" are enshrined in international documents (UDHR, 1948), yet they are currently in the process of being denied for those whose long-term survival is threatened by climate change, materialized not only in deteriorating ecological conditions but also in the non-recognition of their human rights status when they are forced to migrate (Skillington, 2015).

Isaiah Berlin also identified *positive freedom*, defined as involving "the answer to the question 'What, or who, is the source of control or interference that can determine someone to do, or be, this rather than that?'" (Berlin, 1969). For him, positive freedom derives from the desire of a rational individual to be "his own master" able and *free* to achieve self-control, or self-realization. One might enjoy this type of positive freedom when one is "conscious of myself as a thinking, willing, active being, bearing responsibility for his choices and able to explain them by reference to his own ideas and purposes" (Berlin, 1969). Berlin did warn about the potential that positive liberty may lead to "despotism" or to "brutal tyranny", when the "positively free self" is "inflated into some super personal entity – a state, a class, a nation, or the march of history itself" (Berlin, 1969). This collective entity may be promoting "a contentious ideal of the good life behind the veneer of liberty" that marks "a treacherous tilt toward the justification of centralized power under the guise of moral superiority" (Christman, 2005); for this reason, Berlin preferred negative to positive liberty, as negative freedom was "a truer and more humane ideal than the goals of those who seek in the great, disciplined, authoritarian structures the ideal of 'positive' self-mastery by classes, or peoples, or the whole of mankind" (Berlin, 1969). Berlin believed that personal liberty was fundamental in democratically deciding, through "ongoing debate and contestation" (Christman, 2005), what the best life is. He worried about the potential threat to democracy that positive freedom represented, noting that "the connection between democracy and individual liberty is a good deal more tenuous than it seemed to many advocates of both" (Berlin, 1969). The contradictory relationship between individual freedom and democracy had been identified as early as the nineteenth century by Alexis de Tocqueville during his study of democracy in America. Tocqueville explained that while democracy tends to encourage individual expression, there is also the threat posed by "the equalizing tendencies of the democratic system" to individual freedom, when society's interests seem to be predominant over individual rights (Huffman, 1977).

Berlin's analysis of the two concepts of liberty is still considered a piece of "classic" literature, valuable mainly for explaining freedom as an intrinsic feature of human beings. Berlin believed that "to block before a man every door but one, no matter how noble the prospect upon which it opens [...] is to sin against the truth that he is a man, a being with a life of his own"(Berlin, 1969). His concept of positive freedom, understood as self-realization, has been defended by philosophers and historians who either criticized negative freedom as non-intervention (Taylor, 1979; Pettit, 2011), or focused on the quality and effectiveness of human agents as idealized models of power and abilities to act in a certain direction, and not only on human beings' opportunity to act. Skinner, (2008), Pettit (2011) or Christman (2005) believed that "a just society must protect or promote freedom construed in this positive way and see as an ideal the ability of citizens to act as authentic, self-governing agents (that is to be self-realized)" (Christman, 2005). For Christman, the ideal human agents are "autonomous" individuals having the "capacity for critical self-reflection in the development of value systems and plans for action". He believed that adequate resources and just institutions are needed for implementing positive freedom in a society, as "such capacities do not merely emerge naturally, but must be developed through various processes involving educational, social and personal resources" (Christman, 2005).

Freedom protection and erosion in liberalism

The protection of individual freedom is the central tenet of both classical liberalism, the politico-philosophical doctrine born in Great Britain in the mid-seventeenth century, and of modern liberalism extended since the nineteenth century in Western Europe and North America. In both conceptions, the role of state authority was to protect individual freedom from outside interference and to promote it by providing public services such as national defence, law enforcement, sanitation, education, and health care. However, in both conceptions of liberalism, the state power should be limited so as not to become a threat to individual liberty. To prevent the sovereign state from becoming too powerful, two tools could be used: a complex form of balanced government in which power was divided between the monarch, ministers, and Parliament (Britannica, 2023), as well as a hypothetical social contract, whereby the rational, free, and equal individuals assent to institute or to limit the power of the sovereign in exchange for social peace. In the *Leviathan* (1651), Thomas Hobbes described such a social contract in which the governed accepted "the absolute power of the sovereign" in exchange for guaranteed peace and security, or protection against "a return to a state of nature" characterized as the "war of all against all" (Hobbes, 1968). Hobbes defined individual freedom as a natural right (*Jus Naturale*) to the

> Liberty each man hath, to use his own power, as he will himself, for the preservation of his own Nature; that is to say, of his own Life; and consequently, of doing any thing, which in his own Judgement and Reason, he shall conceive to be the aptest means thereunto.
>
> (Hobbes, 1968)

John Locke, the father of modern liberalism, in his *Two Treatises of Government* (1690), accepted the need of a social contract, as he believed that any government is established to protect the people's rights to life, liberty, and property, which are natural rights with "a foundation independent of the laws of any particular society" (Tuckness, 2020). But Locke argued against the subordination of free and equal people to the *absolute power* of a government. He even conceived that free people have the right to resist a government which fails to protect individual rights and provide common goods, and thus provided a justification for the French and the American revolutions (Curran, 2013).

Adam Smith's book *The Wealth of Nations* offers a complex analysis of how human freedom is to be understood in a nation aiming to maximize its wealth in an "obvious and simple system of natural liberty". Smith, considered the father of economic liberalism, believed that a nation-state will be wealthy when its members are enjoying freedom and security to work, save, and invest, as well as trade in an environment of unrestrained competition. His explanation for this outcome was that when individuals are free to promote their own interests, they will also promote the interest of the whole society, through the magic of "the invisible hand", without the need for government intervention.

> Every man [...] is left perfectly free to pursue his own interest in his own way [...]. The sovereign is completely discharged from a duty [for which] no human wisdom or knowledge could ever be sufficient; the duty of superintending the industry of private people, and of directing it towards the employments most suitable to the interest of the society.
>
> (Smith, 1976)

Smith also believed that, by their nature, human beings are equipped with traits that enable them to participate in a social order that is sustainable. One of these traits is "the propensity to truck, barter, and exchange one thing for another", a propensity able not only to produce wealth, through the division of labour, but also to create and sustain "the stock of moral capital" (Rosenberg, 1990), by allowing "people to be increasingly useful to each other" (Lewis, 2000).

While being a promoter of a *laissez-faire, laissez passer* policy, Smith also believed in the rule of law, which was to be enforced by the government, whose main tasks were, besides maintaining order, securing national defence, building infrastructure, and supporting education. These tasks could be carried out, in Smith's view, by a small-size government:

> Little else is requisite to carry a state to the highest degree of opulence from the lowest barbarism, but peace, easy taxes, and a tolerable administration of justice: all the rest being brought about by the natural course of things.
>
> (Stewart, 1793)

Previous to *The Wealth of Nations*, Adam Smith wrote *The Theory of Moral Sentiments* (1759), in which he developed a whole moral theory, as the foundation of both his jurisprudence and his political economy. Smith believed in the existence of natural laws of both physical and moral phenomena, laws which could be discovered and explained. Thus, empirically, mainly through the act of introspection, we know that people have "passions". These may be social (e.g. generosity, compassion, benevolence), antisocial (e.g. hate, envy, revenge), or selfish (e.g. grief, joy, pain, pleasure, self-preservation). How do we channel these passions into a virtuous character which is the "natural object of esteem, honour and approbation", and the ultimate cause to making a society sustainable? While Smith identified this ultimate cause, he explained it in terms of efficient causes. The two efficient causes, acting like checks on people's passions, are the innate principle of "sympathy" and the fictitious character of the "impartial spectator". Sympathy – in Smith's view – consists in fellow-feeling with any actions, passions, and experiences of others, and is the basis of moral approbation.

> How selfish soever man may be supposed, there are evidently some principles in his nature, which interest him in the fortune of others, and render their happiness necessary to him, though he derives nothing from it except the pleasure of seeing it.
>
> (Smith, 1880)

As it creates interdependence, sympathy can be socially constructive. Smith's second efficient cause, the "impartial spectator", is an unbiased observer of how passions are being lived not only by others but by the individual himself. Smith has developed this character in two sub-versions, the "actual spectator" and the "ideal spectator". The actual spectator is any "normal" or "average" person who can place themselves in their imagination into the position of another, feeling what the other must feel. Smith believed that the everyday interaction among actual spectators brings about the rules of justice, which are only imperfect attempts to discover and implement the natural principles of justice: "Our continual observation upon the conduct of others, insensibly leads us to form to ourselves certain general rules concerning what is fit and proper either to be done or to be avoided" (Smith, 1880). The "ideal" spectator "of our sentiments and passions" (Smith, 1880) is the standard of the conscience which also possesses internal knowledge of motives and can judge the sincerity of actions. It is a "much higher tribunal" and has the role of correcting the rules of justice which may be based on mere social opinion. It also acts to point humans to a higher morality:

> Since these, therefore, were plainly intended to be the governing principles of human nature, the rules which they prescribe, are to be regarded as the commands and laws of the Deity, promulgated by those viceregents which he has set up within us [...] those viceregents of God within us, never fail to punish the violation of them, by the torments of inward shame, and self-condemnation; and on the contrary always reward obedience with tranquility of mind, with contentment, and self-satisfaction.
>
> (Smith, 1880)

By imagining the ideal impartial spectator as an institution in itself, Smith actually showed his belief in the existence of external, objective, absolute moral standards which should set the goals of any social morality. "How vain, how absurd would it be for man [...] not to reverence the precepts that were prescribed to him by the infinite goodness of his Creator, even tho' no punishment was to follow their violation" (Smith, 1880).

Possible causes

How, then, did we end up with the pitiful picture of the "economic man" (*Homo economicus*) representing the free person that liberalism, presumably, as championed by Adam Smith, aimed to protect? The "economic man", described as a self-interested "money-making animal" (Ingram, 1888) and which to this day epitomizes a rational economic agent, is not free. He is busy maximizing profits and utility, and continuously growing the economy by participating in more and more intensive (global) trade exchanges. I propose three evolutions which have contributed to this loss of personal freedom at the level of the individual in the Western liberal democracies. First, this is due to efforts made at the end of the eighteenth century and in the nineteenth century by various philosophers (J.S. Mill, J. Bentham, and V. Pareto) to make political economy into an abstract science, in which the "egoism" of the economic agent is elevated to "paradigmatic status" (Fourcade and Healy, 2007). Second, there was the "liberal dream of the civilizing market", which dominated in the eighteenth century, of seeing markets as a civilizing force in society. This dream was challenged by the end of the nineteenth century, when a more complex conception about the role of markets in the society evolved. In 1982, Hirschman developed a typology of markets which included three categories, "civilizing, destructive and feeble" (fragile) markets, based on their impact on society (Hirschman, 1982). Today, free markets have evolved into market societies dominated by neoliberal principles trumping personal liberties under the trio nightmare of commodification, coercion, and exclusion. Third, individual freedom has been and is being challenged by technological advances, the latest of which are in the form of artificial intelligence (AI) and artificial general intelligence (AGI), or generative large language models with their potential to make humans as we know them expendable. These large language models, we are told, "are in the process of birthing an animate intelligence on the cusp of sparking an evolutionary leap for our species" when properly trained "on everything that we humans have written, said and represented visually" (Klein, 2023).

Homo economicus

The hypothetical subject of *Homo economicus* was popularized by John Stuart Mill, who first described this character in his essay "On the Definition of Political Economy; and on the Method of Investigation Proper to It" (Mill, 1836). According to Mill, the economic man's narrow and well-defined motives were a useful abstraction for economic analysis:

> Political economy [...] does not treat of the whole of man's nature as modified by the social state, nor of the whole conduct of man in society. It is concerned with him solely as a being who desires to possess wealth, and who is capable of judging the comparative efficacy of means for obtaining that end.
>
> (Mill, 1836)

Besides a desire to possess wealth and the knowledge of how to accumulate it, Mill's economic man was also seen as a lazy, self-interested individual, more interested in "leisure, luxury and procreation" (Persky, 1995), as well as in invariably doing "that by which he may obtain the greatest amount of necessaries, conveniences, and luxuries, with the smallest quantity of labour and physical self-denial with which they can be obtained in the existing state of knowledge" (Mill, 1836).

Mill was a follower of Jeremy Bentham, the English empiricist and rationalist philosopher who, in the eighteenth century, invented a new social science of human behaviour based on the principle of utility. Bentham believed that he could scientifically describe human nature which, in his view, was ruled by "two sovereign masters, pain and pleasure" (Bentham, 1781/2000). In his conception, these selfish passions not only guide human actions, but also motivate individual human being's search for the good or happiness, defined as utility maximization. It was Bentham's fundamental axiom that achieving "the greatest happiness of the greatest number is the measure of right and wrong" (Bentham, 1781/2000) in any society. In his utilitarian science of human nature, Bentham saw humans not as seekers of freedom but as self-interested seekers of happiness and social security, something that could be achieved not through social relationships, but through coercion of law and social conformity. Bentham was a promoter of negative liberty, as he saw the coercive law as the tool for protection of everyone's individual interest, even if it diminished basic human freedom. Thus, "Bentham's desire to make men happy effectively and consciously de-emphasized the importance of making men free" (Long, 1977). In his writings, Bentham discussed several civil liberties, namely the freedom of the press, the freedom of contracts and competition, and the right of citizens to criticize the government. Yet, these civil liberties were seen from a utilitarian perspective: they were important and deserved protection not for the sake of promoting individual freedom, but because they had the potential to increase the happiness of the greatest number in society. With this "unrealistic and unattractive" understanding of human liberty, Bentham has been considered a "behavioural scientist" (Long, 1977). His utilitarianism introduced "moral individualism" in the history of ideas, a legacy still persisting in the science of economics, where "the individual human being is conceived as the source of values and as himself the supreme value" (Duncan and Gray, 1979), free to make choices based on personal preferences and their abilities.

The shift toward this immoral *Homo economicus* was completed by Vilfredo Pareto, a mathematician-turned-economist, in 1909, when he distinguished between the morals of a person acting in social life and the self-interested attitude of the same person when acting as an economic agent:

> A well-bred man enters a drawing-room: he removes his hat, utters certain words and makes certain gestures [...]. He will justify very many other, different actions, by saying that 'morality requires it'. But let us suppose that this same individual is in the office, engaged in buying [...] a large quantity of grain [...] His purchase of grain will be the last stage in a process of logical reasoning based on certain data of experience.
>
> (Pareto, 1976)

The rise of the market society

The principle of individual freedom was tested in the nineteenth century, as seen through the process of growth in the market society, a process which "demanded a ruthless abnegation of the social status of the human being." (Polanyi, 1944). In his book *The Great Transformation*, Karl Polanyi used empirical social historical data to analyze the causes of the demise of the nineteenth-century European civilization. His conclusion was that:

> The congenital weakness of nineteenth-century society was not that it was industrial but that it was a market society. Industrial civilization will continue to exist when the utopian experiment of a self-regulating market will be no more than a memory.
>
> (Polanyi, 1944)

Polanyi was specifically referring to the episode in Great Britain's economic history when the labour market was created by legislative intervention through the Poor Law Amendment of 1834, an act which cruelly altered the social stratification in the country, by commodifying previously free individuals.

> The New Poor Law abolished the general category of *the poor* [...]. The former *poor* were now divided into physically helpless paupers whose place was in the workhouse, and independent workers who earned their living by labouring for wages. This created an entirely new category of the poor, the unemployed, who made their appearance on the social scene. While the pauper, for the sake of humanity, should be relieved, the unemployed, for the sake of industry, should *not* be relieved. That the unemployed worker was innocent of his fate did not matter.
>
> (Polanyi, 1944)

This British social policy created a self-regulating labour market, and justified more government intervention to protect those "who were now caged in the confines of the labour market" (Polanyi, 1944). Polanyi identified the dynamics of the modern market society as a "double movement" made up of two organizing principles of society:

> The one was the principle of economic liberalism, aiming at the establishment of a self-regulating market, relying on the support of trading classes, and using largely *laissez-faire* and free trade as its methods; the other was the principle of social protection, aiming at the conservation of man and nature, as well as productive organization, relying on the varying support of those most immediately affected by the deleterious action of the market [...] and using protective legislation, restrictive associations, and other instruments of intervention as its methods.
>
> (Polanyi, 1944)

Polanyi considered this "double movement" as a social project which aimed not to secure the freedom of the individual but to implement Bentham's goal to achieve "the greatest happiness of the greatest number" in a society based on economic liberalism. However, Polanyi warned that "Neither freedom nor peace could be institutionalized under that economy, since its purpose was to create profits and welfare, not peace and freedom" (Polanyi, 1944). Polanyi's insights are still valid today when neoliberalist principles have engulfed the whole Global North world. In some countries we can see:

> a spectacular rise in the number of people being put behind bars as the state relies increasingly on police and penal institutions to contain the social disorders produced by mass unemployment, the imposition of precarious wage work and the shrinking of social protection.
>
> (Wacquant, 2001)

Personal freedom is threatened in market societies by drug addictions worsened by drug liberalization and poor market control. This erosion of personal liberty can have long-term consequences over generations, who will be faced with the double task of restoring liberty "in and through political institutions" and of combatting "the malaise of a society that has come to accept an impoverished understanding of it" (Shakelford and Dimsdale, 2023).

The AI revolution

A strong assumption of the sustainability order blueprint, offered to a mankind made up of intelligent and curious terrestrial creatures inhabiting a life-friendly part of a comprehensible Universe, was that humans will eventually develop an increasingly advanced technological civilization. Indeed, over centuries, *Homo sapiens* have discovered the secrets of fire-making, and developed metallurgy, agriculture, and various industries; they have built tools, machines, devices, and infrastructures that freed them from cold, hunger, fear, and heavy burden-carrying, and allowed them to travel farther and learn more details about the natural macro- and microcosm (Denton, 2022). These advances culminated in the twentieth century with the development of the computer and the increase in information creation, data storage capacity, and digitalization processes, or "the integration of intelligent data into everything that we do" (Reinsel et al., 2018). According to a narrative proposed by groups such as the International Data Corporation (IDC), humans are on a journey to create "a data-driven world", or a world run by a huge global database, created by humans' desire and action to access and exchange information. The IDC predicts that the Global Datasphere (the space where data is created, captured, and replicated) will grow from 33 Zettabytes (ZB) in 2018 to 175 ZB in 2025 (Reinsel et al., 2018), where a ZB is equivalent to a trillion gigabytes. However, the IDC also acknowledges that this data-driven world "will be always on, always tracking, always monitoring, always listening, and always watching – because it will be always learning" (Reinsel et al., 2018). This is a scary perspective from a personal privacy and freedom point of view, as humans are faced with the loss of privacy at the hands of the global corporations managing the Global Datasphere. In 2019, Shoshana Zuboff, professor emerita at Harvard Business School, warned that the continuous advances of the digital revolution can compromise human autonomy and democracy. She defined "surveillance capitalism" as the effort by big tech corporations to not only mine our minds for a profit but also to change our behaviour by using AI algorithms. In her words:

> Surveillance capitalism unilaterally claims human experience as free raw material for translation into behavioral data. Although some of these data are applied to product or service improvement, the rest are declared as a proprietary behavioural surplus, fed into advanced manufacturing processes known as "machine intelligence", and fabricated into *prediction*

products that anticipate what you will do now, soon, and later. Finally, these prediction products are traded into a new kind of marketplace for behavioral predictions.

(Zuboff, 2019)

Naomi Klein has called the process of the mining of human minds carried on by the wealthiest companies in history (Microsoft, Apple, Google, Meta, Amazon...) "the largest and most consequential theft in human history" (Klein, 2023), as these companies are

> unilaterally seizing the sum total of human knowledge that exists in digital, scrapable form and walling it off inside proprietary products, many of which will take direct aim at the humans whose lifetime of labor trained the machines without giving permission or consent.
>
> (Klein, 2023)

It is true that these surveillance practices have appeared and intensified during the neoliberal phase of capitalism, but they are not specific to capitalist societies; they are the result of fast, unregulated, and out-of-control information digitization and digitalization processes which have permeated every aspect of ordinary life globally. Authoritarian regimes, such as China's, are also using AI facial recognition and a "social credit system" as instruments of population behaviour control, all in the name of social peace-keeping. In India, the Aadhaar program of digital identification is praised for its potential to make document management, digital payments, and other services very efficient. The program has enrolled close to 1.2 billion people over eight years (more than 85% of India's population) and provides cardless digital IDs which can be used through an open platform called the India Stack. It is hoped that the Aadhaar program will help India implement the Sustainable Development Goals (Gelb and Diofasi Metz, 2017). The latest efforts by AI researchers aim to develop and train "thinking machines" such as large language models (LLMs) – the latest and most famous version of which is GPT-4, developed by OpenAI. The goal is to achieve "artificial general intelligence", defined as AI systems able to "duplicate human intelligence in silicon" (Marks, 2022). The effort is not new; it started in the 1830s, when the English polymath Charles Babbage invented the digital mechanical calculator called by his contemporaries a "thinking machine" (Press, 2023), and which was perfected inside the high-performance "calculators" we are using today, as well as in the numerous miniaturized devices – hearing aids, pacemakers, or insulin detectors – designed to enhance human performance. In parallel, a "science fiction" or speculative fiction type literature was created, envisioning humans' evolution as human-machine, eventually able to achieve immortality (Harari, 2016; Kurzweil, 2005). Are these evolutions simply technological tools or are they dangerous for their ability to curtail humans' freedom? It depends on how we define these new technological realities emerging at a very rapid pace. A definition of AGI on the website of OpenAI simply states that "artificial general intelligence" are AI systems that are "generally smarter than humans" and whose mission is to benefit "all of humanity" by being able to "help us elevate humanity by increasing abundance, turbocharging the global economy, and aiding in the discovery of new scientific knowledge that changes the limits of possibility" (Altman, 2023). However, the CEO of OpenAI warns that "AGI would also come with serious risk of misuse, drastic accidents, and societal disruption" (Altman, 2023). These concerns are real, as in exchange for building machines that supposedly work for humans, the tech companies are requiring humans to "adapt" to the dystopian and unsafe AI society they are creating for them. In a recent expert survey about the future of "high-level machine intelligence" (HLMI), in a question about the probability of "human inability to control future advanced AI systems causing human extinction or similarly permanent and severe disempowerment of the human species", 48 per cent of respondents gave at least 10 per cent chance of an extremely bad outcome (e.g. human extinction) (ESPAI, 2022). Then, we should wonder, why accept all these existential risks in exchange for some potential unverified benefits? These smart, super-intelligent machines built by IT technologists may end up having a "thinking" brain running on data and algorithms, but they will never have a mind, or be able to reason, comprehend, and feel emotions like humans do. We should not count on them to solve our existential problems of climate change, biodiversity loss, poverty, and systematic destruction of our life-support system. I argue that a wise attitude would be for *Homo sapiens* to stop indulging in the "hallucinations" in which they have "been indulging since the rise of modern science and technology especially in hallucinations about man being a God-like creator" (Press, 2023), and start looking for feasible solutions provided by safe human intelligence. In the same way that wisdom prevailed when humans made a decision concerning the prohibition by international law of nuclear weapons, with a treaty that has been in effect since January 22, 2021 (TPNW, 2021), we can hope it will prevail in dealing with the control of risky emerging technologies.

Sustainability as freedom

In 1999, in his book *Development as Freedom*, Nobel laureate economist Amartya Sen conceptualized development as increased human freedom defined as both "capabilities" ("the freedom to achieve various lifestyles") and "functionings" ("the various things a person may value doing or being") (Sen, 1999). While the "capabilities" approach led to the creation of a very useful Human Development Index published since 1990 by the Human Development Report office of the United Nations Development Programme (UNDP), Sen was criticized for not including sustainability in his development approach (Nussbaum, 2003; Crabtree, 2012). Indeed, Sen's concept of positive freedom conceiving people as free to choose the "lifestyles" or the "things" they "may value" has the potential to lead to unsustainable choices.

When the Ecological Footprint indicator, used as a proxy for sustainability, was added to the 2010 Inequality Adjusted Human Development Index (IHDI), countries with a very high IHDI (Norway, USA, Russian Federation) showed levels of very high unsustainability (Crabtree, 2012). However, Sen's analysis of the living human beings, able to visualize what they can do and be when they enjoy both their "capabilities" and their "functionings", is important for its potential to achieve sustainability. The narrative of sustainability as a free gift meant to keep people alive and socially content may be easier borne in the mind of a person who is not exclusively concerned with maximizing their wealth in the form of profits and utility. Real positive freedom can help such persons better control their desires and make purposeful choices for a sustainable lifestyle not based on human preferences which are numerous and subject to changing moods, but on some firm commitments to obey legitimate ecological and moral limits, which can lead to a free and fruitful life.

If we define freedom as "effective personal choice" (Perry, 1940), both "unhampered" (negative) freedom and "implemented" (positive) freedom are relevant for achieving sustainability. Assuming that the desire to use the life-support system, or to live, characterizes all human beings, negative liberty means the removal or absence of those obstacles or constraints (external or internal) on personal decision-making in human life matters. Aspects of negative liberty are freedom from other humans' or government's intervention in issues related to a person's life, health, family life, education, beliefs, expression, etc. External obstacles are laws and regulations that limit personal freedom. They act only for protecting other freedoms considered basic both for the individual and for society. For instance, environmental regulations limiting the right to pollute the ocean aim to protect the health of all ocean life and the freedom of everyone to enjoy the ocean and the ecological services the ocean provides. Negative liberty does not exclude legitimate government intervention in order to limit individual freedom that is hurtful for others. Thus, liberty is "merely that portion of possible choices of action which the state will protect one's asserted right to pursue. It is inevitably a problem of finding a distribution of liberties which will be of maximum benefit to individuals and to society" (Huffman, 1977). Internal obstacles are moral constraints, such as when you realize that "I cannot do this, as it does not seem morally right, or my conscience does not allow me to do it". Positive liberty "means that the externally unimpeded interest is capable of proceeding toward its realization" (Perry, 1940), provided that the necessary resources exist. For instance, a person wishing to install solar panels on their roof can do so, if they know about solar panels, if they can hire someone to install them, and if they have the funds for the operation.

While negative liberty implies less government intervention, positive liberty implies more government intervention (for aiding by providing the necessary resources). Assuming that economic activity is a common-sense requirement for providing human livelihood, Frank Knight considered that any economic activity is a "problem-solving activity" (Knight, 1941) involving human choice. He provided an explanation of the mechanism of human choice-making. The activity starts in the mind and takes place at two levels, at a factual level (what an individual wants) and at a normative level, involving values sorting and clarification (what the individual "ought" to want, given the circumstances). In Knight's words:

> Any statement, uttered or thought, distinguishes between a *state of mind*, which is always in "fact" motivated in some way in the individual mind, and *valuation* in a superindividual or "objective" meaning. Discussion (or even conversation) is not involved either in the phenomenal process or in propositions running in terms of "I want" or "I think" without an implication of "rightness" in the one case and "truth" in the other.
>
> (Knight, 1941)

In this deliberation process, the individual is guided by various values, such as beauty, truth, utility, goodness. Knight considered that truth is the indispensable value for establishing objectivity of a "right" result, which is beyond any individual's subjective desire. Knight defined truth as "a social category, tested and known through agreement with other observers, or judges, and arrived at by the social and intellectual co-operative activity of discussion. This process is indispensable to the very idea of objectivity" (Knight, 1941). C.S. Lewis also underlined the importance of objective values in making decisions in a democratic context, as "subjectivism about values is eternally incompatible with democracy" (Lewis, 1943). He insisted that "a dogmatic belief in objective value is necessary to the very idea of a rule which is not tyranny or an obedience which is not slavery" (Lewis, 2002). Lewis considered that the existence of an objective moral law was a necessary condition to avoid losing one's individual freedom under an authoritarian, fascist, or communist regime. "The very idea of freedom presupposes some objective moral law which overarches rulers and ruled alike. [...] We and our rulers are of one kind only so long as we are subject to one law" (Lewis, 1943). I agree with Lewis, and I believe that the fragmented morality that has dominated Western cultures in the last two centuries (MacIntyre, 1981) has undermined human freedom and has made it difficult to establish the truth of what sustainability is and how individuals can implement it. The lack of one objective moral rule has led to creation of numerous relative "moral" rules, and the voluntary subjection of the human selves – "liberated" from the objective norms and rules aiming to defend their primordial "birthright" to live – to poorly understood rules and cruel rulers, interested not in the sustainability of the human race but in maximizing profits or maintaining power. While hoping to reject crude utilitarianism in our daily choices, it is useful to remember Shakespeare's wisdom that humans are unable on their own to always control their passions. They need grace from above. As Shakespeare stated in *Antony and Cleopatra*, "truly human life is sustained by a transcendent cosmic order that gives grace to live nobly and die well. This is why we must renounce raw power and forsake its perverse pleasure" (Ranasinghe, 2022). Only a holistic understanding of freedom, "which rests upon the integrity of life, not in the mind alone, not in the will alone, not in activity alone, but in the whole man" (Cobb, 1941), can restore the ability of human beings to correctly understand sustainability and freely choose to live according to its normative rules. These rules can protect humans' hope to live beyond their "fate to die" by transforming "a dead yesterday into a living tomorrow" (Cobb, 1941). If we are serious about the Brundtland report's commitment to living in a way which does not "compromise the ability of future generations to meet their own needs", we need to make sure that there will be a next generation of strong and morally upright humans to enjoy the life-sustaining system.

This chapter has discussed the relevance of freedom for understanding and implementing sustainability. Based on Isaiah Berlin's concepts of negative and positive freedom, it has shown that while freedom is the objective birthright of any human being, the choice of a sustainable lifestyle depends on understanding and acting upon both aspects of liberty, negative and positive, which sometimes require deliberate sacrifices of freedom. It has analyzed the history of liberty protection under the classical and the modern liberalist doctrines, the birth of the utilitarian doctrine of moral individualism which instituted freedom as preference satisfaction, and has identified three causes of erosion of individual freedom under modern liberalism: seeing humans as *Homo economicus*, the creation of a free-market society, and the fast process of digitalization culminating in the efforts to achieve AGI systems. Finally, it has suggested ways to make sustainable choices by a free *Homo sapiens* empowered with "capabilities" and "functionings", and able to control their desires through a deliberation process guided by objective values. Achieving sustainability beyond "the humans' fate to die", requires deliberately "transforming a dead yesterday into a living tomorrow" by preparing today the next generation of sustainable inhabitants of planet Earth.

References

Altman, S. (2023) "Planning for AGI and beyond." OpenAI website: https://openai.com/blog/planning-for-agi-and-beyond.

Bentham, J. (2000) *An Introduction to the Principles of Morals and Legislation.* Kitchener, ON: Batoche Books. (Original work published 1781)

Berlin, I. (1969) "Two Concepts of Liberty." In *Four Essays on Liberty.* Oxford: Oxford University Press, pp. 118–172.

Britannica (2023) "Classical liberalism." www.britannica.com/topic/classical-liberalism.

Cambridge (2020) *Advanced Learner's Dictionary & Thesaurus.* Cambridge: Cambridge University Press.

Chesterton, G.K. (1969) *Autobiography*, introduction by A. Burgess. London: Hutchinson.

Christman, J. (2005) "Saving positive freedom." *Political Theory*, 33(1): 79–88.

Cobb, H.V. (1941) "Hope, Fate, and Freedom: A Soliloquy." *Ethics*, 52(1): 1–16.

Crabtree, A. (2012) "A Legitimate Freedom Approach to Sustainability: Sen, Scanlon and the Inadequacy of the Human Development Index." *The International Journal of Social Quality*, 2(1): 24–40.

Curran, E. (2013) "An Immodest Proposal: Hobbes Rather Than Locke Provides a Forerunner for Modern Rights Theory." *Law and Philosophy*. 32: 515–538.

Daly, H. and Farley, J. (2011) *Ecological Economics: Principles and Applications*. Washington, DC: Island Press.

Denton, M. (2022) *The Miracle of Man: The Fine-Tuning of Nature for Human Existence*. Seattle, WA: Discovery Institute Press.

Duncan, G. and Gray, J. (1979) "The Left Against Mill." In *New Essays on John Stuart Mill and Utilitarianism*, edited by W.E. Cooper, K. Nielsen, and S.C. Patten. Guelph, ON: Canadian Association for Publishing in Philosophy.

Eckersley, R. (2020) "Ecological democracy and the rise and decline of liberal democracy: looking back, looking forward." *Environmental Politics*, 29(2): 214–234.

Ehrlich, P.R. (2000) *Human Natures, Genes, Cultures, and the Human Prospect*. Washington, DC: Island Press.

ESPAI (2022) "Expert Survey on Progress in AI". AI Impacts. https://aiimpacts.org/2022-expert-survey-on-progress-in-ai.

FIW (2023) *Freedom in the World 2023*. Freedom House. https://freedomhouse.org/sites/default/files/2023-03/FIW_World_2023_DigtalPDF.pdf.

Fourcade, M. and Healy, K. (2007) "Moral Views of Market Society." *Annual Review of Sociology*, 33: 286–311.

Gallie, W. (1955) "Essentially contested concepts." *Proceedings of the Aristotelian Society*, 56, 167–198.

Gelb, A. and Diofasi Metz, A. (2017) "Identification Revolution: Can Digital ID Be Harnessed for Development?" CGD Brief. www.cgdev.org/sites/default/files/identification-revolution-can-digital-id-be-harnessed-development-brief.pdf.

Harari, Y.N. (2016) *Homo Deus: A Brief History of Tomorrow*. New York: Random House.

Hirschman, A.O. (1982) "Rival interpretations of market society: civilizing, destructive, or feeble?" *Journal of Economic Literature*, 20: 1463–1484.

Hobbes, T. (1968) *Leviathan*, edited by C.B. Macpherson. London: Penguin Books. (Original work published 1651)

Hueting, R. and Reijnders, L. (1998) "Sustainability as an Objective Concept." *Ecological Economics*, 27(2): 139–147.

Huffman, J.L. (1977) "Individual Liberty and Environmental Regulation: Can We Protect People While Preserving the Environment?" *Environmental Law*, 7(3): 431–447.

Ingram, J.K. (1888) *A History of Political Economy*. Edinburgh: Adam and Charles Black.

Kane, R. (1996) *The Significance of Free Will*. Oxford: Oxford University Press.

Klein, N. (2023) "AI machines aren't 'hallucinating'. But their makers are." *The Guardian*, 8 May 2023. www.theguardian.com/commentisfree/2023/may/08/ai-machines-hallucinating-naomi-klein.

Knight, F.H. (1941) "The Meaning of Freedom." *Ethics*, 52(1): 86–109.

Kurzweil, R. (2005) *The Singularity is Near: When Humans Transcend Biology*. London: Penguin.

Lewis, C.S. (1943) "The Poison of Subjectivism." In *Christian Reflections*, edited by W. Hooper. Grand Rapids, MI/Cambridge, UK: William B. Eerdmans Publishing Co.

Lewis, C.S. (2002) "The Abolition of Man" in *The Complete C.S. Lewis Signature Classics*. New York: HarperOne. (Work originally published 1944)

Lewis, T.J. (2000) "Persuasion, Domination and Exchange: Adam Smith on the Political Consequences of Markets." *Canadian Journal of Political Science* 33(2): 273–289.

Long, D.G. (1977) *Bentham on Liberty: Jeremy Bentham's Idea of Liberty in Relation to His Utilitarianism*. Toronto and Buffalo: University of Toronto Press.

MacIntyre, A.C. (1981) *After Virtue: A Study in Moral Theory*. Notre Dame, IN: University of Notre Dame Press.

Marks, R.J. (2022) *Non-Computable You. What You Do That Artificial Intelligence Never Will*. Seattle, WA: Discovery Institute Press.

McGinnis, J.O. (2023) "A Child's Primer for Liberty." *Law and Liberty*, April 20, 2023. https://lawliberty.org/a-childs-primer-for-liberty.

Meadows, D.H., Meadows, D.L., Randers, J., Behrens III, W.W. (1972) *Limits to Growth*. Potomac Associates. https://doi.org/10.1349/ddlp.1.

Meadows, D., Randers, J., and Meadows, D. (2004) *The Limits to Growth: The 30-Year Update*. New York: Earthscan.

Mill, J.S. (1836) "On the Definition of Political Economy; And on the Method of Investigation Proper to It. Essay V." In *Essays on Some Unsettled Questions Of Political Economy*. Kindle Edition.

Mill, J.S. (1859) *On Liberty*. The Project Gutenberg eBook of On Liberty, 2011. www.gutenberg.org/ebooks/34901.

Murphy, Jr. T.W., Murphy, D.J., Love, T.F., LeHew, M.L.A., and McCall, B.J. (2021) "Modernity is incompatible with planetary limits: Developing a PLAN for the future." *Energy Research and Social Science*, 81. https://doi.org/10.1016/j.erss.2021.102239.

Nussbaum, M.C. (2003) "Capabilities as Fundamental Entitlements: Sen and Social Justice." *Feminist Economics*, 9(2–3):33–59.

Pareto, V. (1976) *Sociological Writings*, selected and introduced by S.E. Finer. Oxford: Basil Blackwell. (Original work published 1909)

Perry, R.B. (1940) *On Freedom: Its Meaning*, edited by R. Anshen. New York: Harcourt, Brace & Co.

Persky, J. (1995) "Retrospectives: The Ethology of Homo Economicus." *The Journal of Economic Perspectives*, 9(2): 221–231.

Pettit, P. (2011) "The Instability of Freedom as Non-Interference. The Case of Isaiah Berlin." *Ethics*, 121: 693–716.

Polanyi, K. (1944) *The Great Transformation*. Boston, MA: Beacon Press.

Press, G. (2023) "Artificial General Intelligence (AGI) Is A Very Human Hallucination." *Forbes*, March 28, 2023. www.forbes.com/sites/gilpress/2023/03/28/artificial-general-intelligence-agi-is-a-very-human-hallucination/?sh=5ae0458764f2.

Ranasinghe, N. (2022) *Shakespeare's Reformation. Christian Humanism and the Death of God*, edited by L. Oser. South Bend, IN: St. Augustine's Press.

Raworth, K. (2017) *Doughnut Economics: Seven Ways to Think Like a 21st Century Economist*. White River Junction, VT: Chelsea Green Publishing.

Reinsel, D., Gantz, J., and Rydning, J. (2018) "The Digitization of the World from Edge to Core. Data Age 2025." An IDC White Paper – #US44413318. www.seagate.com/files/www-content/our-story/trends/files/idc-seagate-dataage-whitepaper.pdf.

Robinson, J. (2004) "Squaring the circle? Some thoughts on the idea of sustainable development." *Ecological Economics*, 48(4): 369–384.

Rockström, J., Steffen, W., Noone, K., Persson, Å. *et al.* (2009) "Planetary boundaries: Exploring the safe operating space for humanity." *Ecology and Society*, 14(2): 32. www.ecologyandsociety.org/vol14/iss2/art32.

Rosenberg, N. (1990) "Adam Smith and the Stock of Moral Capital." *History of Political Economy*, 22(1): 1–18.

Rousseau, J.J. (2012) *The Social Contract and Other Political Writings*, translated by Q. Hoare, edited by C. Bertram. London: Penguin.

Sellars, R.W. (1946) "Positivism and Materialism." *Philosophy and Phenomenological Research*, 7(1): 12–41. https://doi.org/10.2307/2103022.

Sen, A. (1999) *Development as Freedom*. Oxford: Oxford University Press.

Shakelford, K. and Dimsdale, T. (2023) "On the Need for Scholars and Warriors." Law and Liberty, April 27, 2023. https://lawliberty.org/on-the-need-for-scholars-and-warriors.

Skillington, T. (2015) "Climate justice without freedom: Assessing legal and political responses to climate change and forced migration." *European Journal of Social Theory*, 18(3): 288–307.

Skinner, Q. (2008) "Freedom as the Absence of Arbitrary Power." In *Republicanism and Political Theory*, edited by C. Laborde and J. Maynor. Oxford: Blackwell, pp. 83–101.

Smith, A. (1976) *The Wealth of Nations*, edited by by R.H. Campbell, A.S. Skinner, and W.B. Todd. Oxford: Oxford University Press. (Original work published 1776)

Smith, A. (1880) *The Theory of Moral Sentiments*. London: George Bell.

Stewart, D. (1793) "Account of the Life and Writings of Adam Smith, LL.D." https://delong.typepad.com/files/stewart.pdf.

Steffen, W., Richardson, K., Rockström, J., Cornell, S.E.*et al.* (2015) "Planetary boundaries: Guiding human development on a changing planet." *Science*, 347 (2015): 1259855. https://doi.org/10.1126/science.1259855.

Taylor, C. (1979) "What's Wrong with Negative Liberty." in *The Idea of Freedom*, edited by A. Ryan (ed.), Oxford: Oxford University Press.

TPNW (2021) Treaty on the Prohibition of Nuclear Weapons. Washington, DC: Nuclear Threat Initiative. www.nti.org/education-center/treaties-and-regimes/treaty-on-the-prohibition-of-nuclear-weapons.

Trenchard, J. and Gordon, T. (1971) *Cato's Letters*. New York: Da Capo.

Tuckness, A. (2020) "Locke's Political Philosophy." *The Stanford Encyclopedia of Philosophy* (Winter 2020 Edition), edited by E.N. Zalta. https://plato.stanford.edu/archives/win2020/entries/locke-political.

UDHR (1948) Universal Declaration of Human Rights. UN General Assembly resolution 217A. Paris, 10 December 1948. www.un.org/en/about-us/universal-declaration-of-human-rights.

Wackernagel, M. and Rees, W. (1996) *Our Ecological Footprint: Reducing Human Impact on the Earth*. Philadelphia: New Society Publishers.

Wacquant, L. (2001) "The penalization of poverty and the rise of Neoliberalism." *European Journal of Criminal Policy and Research*, 9: 401–412.

Wilson, E.O. (1998) *Consilience*. New York: Knopf.

Zuboff, S. (2019) *The Age of Surveillance Capitalism: The Fight for a Human Future at the New Frontier of Power*. New York: PublicAffairs.

8   Who is the sustainer?

Sustainability, as I have defined it, is the mysterious existence in the Universe of a web of dynamic interconnections, relationships, and dependencies that work together to support life on Earth, enabled by voluntary and involuntary responses of all the living beings participating in this webwork. The involuntary responses are due to functions and mechanisms inherent in living creatures such as a heartbeat, respiration, and the drive to eat and drink, and to reproduce. Some of these vital processes are easily observable and understandable, while some are hard to explain, such as how do Pacific salmon know in October of each year how to navigate from the waters of the Pacific Ocean back to the rivers where they were born in order to spawn, and afterwards die, in the river, or what force enables house plants in windows to turn toward sunlight. Voluntary responses are those over which humans, as rational and responsible participants in the workings of the life-web, have some control, such as when someone adopts a certain diet or decides how many offspring to produce. But the mechanisms enabling such decisions or choices, as well as the unintended consequences of these decisions, are largely unknown, in spite of remarkable advances in the neurological and physiological sciences. While these sciences have unveiled much information about how the brain functions, how the human mind functions is still a mystery (Jerath and Beveridge, 2018). An honest approach to the sustainability issue cannot evade the mysterious dimension of sustainability, where "mystery" is defined as "an enduring part of the human condition and not just another word for future knowledge not yet discovered" (Daly and Farley, 2011).

There are at least three mystery questions concerning sustainability: (1) What is the origin and structure of the life-supporting system available to all living creatures on planet Earth? (2) How did life originate and become organized in information-storing DNA molecules able to sustain and reproduce life? (3) What explains the mysterious configuration of the human creature, who, according to the Irish philosopher John Scotus, "sums up in himself the material world and the spiritual world" (Copleston, 1993)? Ultimately, the question overarching these questions is whether there is an initiator and sustainer of this amazing living network and who this might be.

DOI: 10.4324/9781003307587-9

There have been numerous attempts to give answers to these ultimate questions, starting with the medieval scholars who used both their religious perspectives and logical reasoning to deal with the "three medieval problems", namely "the eternity of the world; divine omniscience and human freedom; the soul and immortality." (Marenbon, 2023). The answers depended on their worldviews which, at that time, had been influenced by Classical Greek philosophy with its clear division of philosophers into two groups: the *materialists* who believed that matter and energy are the only entities that exist and which can explain everything in the Universe; and the *idealists* who believed in the primordiality of mind over matter and energy. The rise of modern science in the seventeenth century and the spread of the Enlightenment belief, in the eighteenth century, that reason and science can give people more knowledge than tradition and religion, led to a dominant materialist worldview in Western countries during the nineteenth and the twentieth centuries, which has spread into the pervasive secularism of today's modern and postmodern cultures. These cultures, characterized by "post-metaphysical thinking" in science, pluralism, "both positive law and profane morality", and "a naturalistic self-objectification of persons" in everyday life (Habermas, 2006), being based in the belief that "matter – the stuff of physics – underlies everything real" (Axe, 2016), allow people to make unsustainable choices with total impunity. The goal of this chapter is to show that the answer to the unavoidable logical question "Who is the Sustainer?" forces us to probe into a reality beyond that depicted by a pure materialistic worldview, a stratified reality where humans are not only bodies with brains, but also spiritual creatures able to know reality not only phenomenologically and rationally, but also as creatures with intuitions and faith. Here, faith is understood as "a human problem that all of us, whether participants in a religious tradition or not, must confront" (Grant, 1984). Two hypotheses will be developed and considered by using abductive reasoning in the attempt to arrive at the inference to the best explanation shedding light on the question.

A short history of two answers to the deep questions of life on Earth

One of the first *materialist* natural philosophers was the Greek pre-Socratic atomist philosopher Democritus (born about 460 BC). Democritus believed that the building blocks of the physical Universe are the atoms, defined as small indivisible, uncreated, and indestructible particles of matter, that move about in the void. From the collision of atoms, molecules are formed, which have evolved to simple life forms which could then lead to complex life forms including conscious minds. The materialist principles were adopted by the Enlightenment philosophers such as Thomas Hobbes, Auguste Compte, and David Hume, and were widely accepted as the mainstream scientific explanation of the origin of new forms of life, after the publication in 1859 of Charles Darwin's book *On the Origin of Species*. Even if Darwin's book does not explain the origin of the first living cell, by providing a scientific explanation for evolution of new species through a naturalistic mechanism based on

chance and natural selection, the book eliminated the need for divine intervention in accounting for the origin of life. Some philosophers of science
believe that "the God hypothesis" was eliminated from science by the French
scientist Pierre-Simon Laplace at the end of the eighteenth century when he
published a multi-volume *Treatise of Celestial Mechanics* in which he developed a theory of the origin and functioning of the solar system based entirely
on purely natural gravitational forces (Meyer, 2021). According to an anec-
dote recorded in history, "when Laplace offered Napoleon a copy of one of
his books on astronomy, the latter asked: 'I am told that in this great book
you have written on the system of the world there is no mention of God, its
creator', to which Laplace replied: 'Sire, I have had no need of that hypothesis'." (Durán, 2019).

Today, scientific materialism or methodological naturalism are the only
accepted scientific methods providing explanations for theories of origin either
for the Universe or for life, including for rational human beings. For instance,
in 2010, physicists Stephen Hawking and Leonard Mlodinow stated that the
Universe arose spontaneously from nothing, just by following the laws of
nature: "Because there is a law such as gravity, the universe can and will create
itself from nothing [...]. Spontaneous creation is the reason there is something
rather than nothing, why the universe exists, why we exist" (Hawking and
Mlodinow, 2010). As a consequence, in Western liberal democratic societies,
humans are conceived in pure materialistic terms, as purposeless individuals
who act driven by instincts and rational minds, and are destined to extinction
in spite of impressive human achievements. It is a sad existential predicament
for humanity when one believes "that man is the product of causes which had
no prevision of the end they were achieving; that his origin, his growth, his
hopes and his fears, his loves and his beliefs, are but the outcome of accidental
collocations of atoms", as philosopher Bertrand Russell wrote in 1904 (Russell,
1957). Russell's view remains widely shared by scientists and philosophers such
as physicist Lawrence Krauss, who believes that humans are a cosmic accident
in a purposeless Universe, in which we, humans, are expected to "make our
own purpose" and "make our own joy" (Krauss, 2017).

The second group of philosophers, those called *idealists*, believed that ideas
coming from a mind or from beyond the physical world are the "first principles" that create and shape the material world. Ancient Greek philosophers,
such as Aristotle (384–322 BC), believed that life on Earth followed a pattern
"that seemed rational to the human mind" (Hooykaas, 1972), as it was
"issued from an intrinsic self-existent logical principle called the *logos*, rather
than from a mind or a divine being with a will" (Meyer, 2021). According to
this view, life "was always conceived to be an integral part of nature, and its
constituent forms – substantial forms – were basic components of the world-
order" (Denton, 1986). These "substantial forms" were considered to act as
"active agents in nature, molding the form of organisms, and through their
collective activities, the overall pattern of life on earth" (Denton, 1986). For
Aristotle, nature and all living bodies had a "soul", which was the "form"

giving the body its life functions, movement, feelings, and thinking; this meant that the soul would only live while the body was alive. However, Aristotle believed in the eternity of the world which had no beginning and included a system of nature which "has always been as it is, and so it will continue" (Marenbon, 2023). The Aristotelian view of the world's eternity was not accepted by religious believers from the three main monotheistic faith traditions (Jews, Christians, and Muslims) for whom the world had a beginning, being created by the uncreated God who is omniscient and exists in eternity, not in time. Some scholars from these faith traditions also questioned Aristotle's view about the mortality of the soul, as they believed in bodily resurrection, or the immortality of the soul. Both ancient Greek and Roman scholars, such as Aristotle and Ptolemy of Alexandria, agreed on the fact that the Earth is a sphere at the centre of the Universe and that its purpose is to sustain human life. This geocentric model of the Universe was displaced in 1543 by the Polish astronomer Nicolaus Copernicus's heliocentric model which, by affirming that the sun was at the centre of the Universe, discredited the medieval cosmology of unity between man and the cosmos. In Copernicus's model, not only the Earth but also the human being lost the privileged position of being at the centre of the Universe (Koyre, 1957; Denton, 2022). The idea that the Universe was organized according to some rational "order of things", and displayed natural law-like regularities, as well as the understanding that the "order of nature" was comprehensible to humans, either rationally or through empirical investigation, led to the birth of Western modern science (Whitehead, 1925) and the development of the scientific method during the sixteenth and the seventeenth centuries. The scientific method was based on the need to discover the causes of effects seen in nature, by using experimentation, hypothesis development and testing, and inductive approaches helped by technologies. This view was held by natural philosophers such as Francis Bacon (1561–1626) and Robert Boyle (1627–1691), both champions of experimental science, who believed that "the hidden structure of the natural world is too subtle to be penetrated by the Aristotelian, deductive approach to science, and that technology can aid in our investigation of the natural world" (Eaton, n.d.). Robert Boyle, the father of modern chemistry, even used an air pump to create a vacuum, thus rejecting Aristotle's idea that nature has a "soul" and when thus personified it can "abhor a vacuum". Many of the first modern scientists, such as Newton, Galileo, Kepler, Boyle, John Ray, were theists or believers in the Judeo-Christian God of the Bible, and so developed and tested theories in physics, astronomy, chemistry, and biology based on the assumption that the world was created by God and that "the natural order is the product of a single intelligence from which our own intelligence descends" (Fuller, 2007). It was the understanding of these scientists that humans could access information and knowledge about what God actually did in the created world by reading three "books of wisdom": "the book of scripture," "the book of conscience", and "the book of nature" (Eaton, n.d.). Describing nature as a "book" which can be investigated empirically to discover its secrets was just one metaphor used by the first modern natural philosophers acting also as scientists. Two other metaphors used to describe the order and working of the Universe were that of a clockwork (or that of a carefully crafted mechanism), and that of a collection of laws or regularities. For instance, Robert Boyle, considered one of the first "mechanical philosophers", described nature as "a great machine, an intelligently constructed system of unintelligent matter in motion rather than a living organism with a 'soul' or 'intelligence' of its own" (Davis, 2013). Boyle identified design in the Universe which functioned

> like a rare Clock, such as maybe that at *Strasbourg*, where all things are so skillfuly contriv'd, that the Engine being once set a Moving, all things proceed according to the Artificer's first design, [...] and do not require, like those of Puppets, the peculiar interposing of the Artificer, or any intelligent agent employed by him, but perform their functions upon particular occasions, by virtue of the General and Primitive Contrivance of the whole Engine.
>
> (Boyle, 1686)

The third metaphor described nature as a "law-governed realm" (Meyer, 2021), functioning according to immutable "natural regularities" called, since the seventeenth century, "laws of nature". The laws of nature were discovered and not invented, being considered to be "contingent forms of order that were *impressed* on nature from the outside by a creator" (Oakley, 1961). This Creator, according to the astronomer Johannes Kepler, "wanted us to recognize" the natural laws "by creating us after his own image so that we could share in his own thoughts" (Meyer, 2021). Kepler's insights came after he discovered the unexplainable precision of planetary motion that he could express in mathematical formulas. The fact that the laws of physics can be expressed in the precise language of mathematics continues to amaze scientists to this day. "The miracle of the appropriateness of the language of mathematics for the formulation of the laws of physics is a wonderful gift which we neither understand nor deserve", stated, in 1960, the Nobel Prize physicist Eugene Wigner (Wigner, 1960). Isaac Newton also believed that the laws of nature "implied reliable divine oversight" (Meyer, 2021), especially when the respective laws involved mysterious phenomena such as the attraction force in Newton's law of gravitation. Newton described the universal law of gravitation in his book *Philosophiae Naturalis Principia Mathematica*, published in 1687, without providing a mechanistic explanation for the gravitational attraction between material objects situated at a distance. He provided instead a mathematical formula for the law of gravitation which states that two material bodies exert an attraction force on each other proportional to the product of their masses and inversely proportional to the square of the distance between them. When questioned by the polymath Gottfried Leibniz about the cause of the mysterious force of attraction which seemed to act without a material cause at a distance, Newton acknowledged that he did not know what caused gravitational attraction but, in private correspondence, he affirmed his belief "that an *immaterial* agent, God, was responsible for the mysterious action that his law described" (Meyer, 2021).

There is no science–religion conflict

It is now accepted by numerous historians and philosophers of science (Butterfield, 1957; Barbour, 1997; Hodgson, 2001) that the faith in a Creator of the first scientists led to the rise of modern science in the Western civilization in the sixteenth and seventeenth centuries. However, with the advance of scientific materialism, science and theistic faith have been seen as either incompatible or in open conflict (Draper, 1874; Dawkins, 1986; Hawking, 1998). This has led to deliberate efforts to exclude from science any metaphysical explanations, dubbed as pre-scientific theological thinking, or "nonsensical twadle –'sophistry and illusion', as Hume says, which we should 'commit to the flames'" (Popper, 1968), and to proclaim rational facts-based science as the unquestioned arbiter of truth in the public sphere. All this while relegating faith, largely understood as "religion", which Marx had declared in 1843 to be "the opium of the people" (Marx, 1970), to the confines of private life. Scientific discoveries and high rates of economic growth have led, mainly after the Second World War, to a rapid process of modernization and to growing levels of "existential security", which "are also conducive to secularization – a systematic erosion of religious practices, values and beliefs" (Inglehart, 2018), in the liberal democracies of the world. However, around the turn of the twenty-first century, the thesis of the inevitable secularization of modernizing societies, based on the assumption that "in the long run, religious views will inevitably melt under the sun of scientific criticism and that religious communities will not be able to withstand the pressures of some unstoppable cultural and social modernization" (Habermas, 2006), has been challenged by an increase in the social and political importance of religious traditions, movements, and faith communities all over the world (Berger, 1999). This trend started with the 1979 Islamic Revolution in Iran and with publications defining new theoretical concepts, such as "the revival of religion" or "the revenge of God" (Kepel, 1994), or "the clash of civilizations" (Huntington, 1993), or "the de-secularization of the world" (Berger, 1999). The World Values Survey and the European Values Survey carried out between 1981 and 2012, confirmed that due to "widely differing human fertility rates, a larger share of the world's population is religious today than it was 30 years ago" (Inglehart, 2018). The rise in fundamentalist religions in the Middle East, in Africa, in India, and in South-East Asia, and the spread of evangelical Christianity in Latin America, in South Korea, and in the former communist countries of Eastern Europe clearly demonstrate that "religion has not died out and shows no sign of doing so" (Inglehart, 2018). On the contrary, "religion is a permanent feature of human society" (Inglehart, 2018), and has become, since the beginning of the twentieth century, an important factor "in the public sphere" (Habermas, 2006) and in international relations (Haynes, 2005). Used by state and non-state actors, both to "keep societies together and to separate them" (Ozturk, 2023), religion is today widely accepted in the globalized world as a political tool containing specific elements of "soft power" (Nye, 1990), defined as influence exercised by one country over another by using "soft" tools, such as diplomacy, culture, language, or education. The concept of "soft and hard religions" was introduced by Johan Galtung in a 1997–1998 article exploring the peace potential of various religions "in terms of their inclination to condone or reject violence" (Galtung, 1997–1998). Galtung defined theistic religions (Judaism, Christianity, and Islam) based on a unique, transcendental God "who chooses people", as "hard", and the Eastern religions (Hindu and Buddhism), based on numerous gods "chosen by people", as "soft". Religious "softness", as explained by Galtung, consists in underlying the "non-dividing aspects of religion", in seeing God as "immanent inside people", and "perhaps, no longer a subject; more like a substance lifting us", and in giving people more flexibility in their choices of the "sacred". The article concluded that "For the sake of world peace, dialogue within religions and among them must strengthen the softer aspects" (Galtung, 1997–1998) of religions. This is exactly what has happened in the last two decades in the global "soft religion" sphere, where the dividing lines between world religions are more porous, the soft power of religions merges with countries' secular soft power (Ozturk, 2023), or the line between "religion" and "the secular" is increasingly blurred in national politics. This trend can be exemplified by the appearance of a new form of secularist "identitarian Christianism" in Europe, where "Christianity is embraced not as a religion but as a civilizational identity understood in antithetical opposition to Islam" (Brubaker, 2017). In spite of the re-emergence of religions in world politics, international relations still take place mainly in the secular environment instituted by the Westphalia Peace Treaties of 1648. There are no guarantees that the revived "soft religion" power can contribute to world peace or to global sustainability in a world of fragmented moralities (MacIntyre, 1981). The amalgamation of "soft" religion and politics could be an attempt to find the meaning of existence lost when religion was excluded from among the sources of mainstream knowledge and when science was proclaimed as the only way to the truth. In fact, both science and religion are involved in the same human endeavour, namely the search to discover and understand how the world works and what humans' role may be in it. To me, both are important in promoting sustainability as a way of thinking and to give meaning to life on Earth. As Jonathan Sacks said, "Science takes things apart to see how they work. Religion puts things together to see what they mean. They speak different languages and use different powers of the brain" (Sacks, 2011). Research in brain neuroscience has confirmed that humans use the left hemisphere of the brain for "fast" thinking of abstract thoughts, by manipulating black-and-white pieces of information, to gain "apprehension" of the world, while the right hemisphere of the brain is used for "slow" thinking in order to get the whole picture or the "understanding" of complex, evolving, and unpredictable processes characteristic to life (Kahneman, 2011; McGilchrist, 2021). These results of neuroscience research show that Alfred North Whitehead was right when he assessed that "the antagonism between science and metaphysics has, like all family quarrels, been disastrous" (McGilchrist, 2021) for Western societies. Excluding metaphysics from science has made science "conspicuously successful for understanding and manipulating the material universe" (Dyson, 2006) by humans, but has led to the loss of an important source of meaning and identity for them, as religion can give us "hints of a mental or spiritual universe that transcends the material universe" (Dyson, 2006). It is important to remember Nobel laureate biologist Peter Medawar's assessment that, in spite of all the wonderful achievements of science, science has inherent limits, given

> the existence of questions that science cannot answer and that no conceivable advance of science would empower it to answer. These are the questions that children ask – the "ultimate questions" of Karl Popper [...] How did everything begin? What are we all here for? What is the point of living?
>
> (Medawar, 1984)

Medawar's answer was: "It is not to science, therefore, but to metaphysics, imaginative literature or religion that we must turn for answers to questions having to do with first and last things" (Medawar, 1984). While it is true that "science cannot make any pronouncements about ethical principles", it does not mean "that there are no such principles while in fact the search for truth presupposes ethics" (Popper, 1978). If science is the process of conscious exploring in the search for truth, it is the duty of any scientist to search for explanations, even metaphysical explanations, to any questions, including ultimate questions, and to follow the evidence where it leads.

Using science to search for the Sustainer

Science cannot tell us much about a miraculous Sustainer of the life-supporting system, as there are no scientific experiments that can be conducted to detect such an entity. David Hume stated long ago that humans do not have previous "experience of the origin of worlds" (Hume, 1989). However, science can shed light on an undeniable fact, namely that the life-sustaining mechanism is a marvel of creativity, as seen in the physical nature, in the living world, and in the miraculous way in which creative minds have produced, throughout our somewhat short humankind history, art, music, poetry, scientific discoveries, inventions, and technical innovations. The creativity seen in nature does not just give the illusion of creative design and sustainability, it speaks about real design and ability to sustain life and its conditions. An example within humans' experience, for instance, is the wonder of seeing babies grow through various stages not only in their physical and metabolical development but also in their understanding of the world around them. This human ability to intuit design in the world humans see has been called the "universal design intuition" by molecular biologist Douglas Axe (2016). The design intuition manifests itself from early ages in children who, by age 4 or 5, according to psychologist Alison Gopnik, "start to invoke an ultimate God-like designer to explain the complexity of the world around them – even children brought up as atheists" (Gopnik, 2014). The design intuition enlarges the field of knowledge sources by including among them imagination, faith and "heart-felt knowledge", as when someone says, "I know in my heart that this must be true." If we accept, from first-hand experience, that the design intuition is real, then the logical law of causality, which underlines any scientific inquiry, forces us to look for an author and a sustainer of our miraculous life-sustaining Universe and everything in it. From Kant, we know that "everything that happens presupposes a previous condition, which it follows with absolute certainty, in conformity with a rule [...] All changes take place according to the law of the connection of Cause and Effect" (Kant, 1878/1781). A good question is then whether we find evidence about what or who is behind this carefully crafted life-support system that our intuition tells us is designed? I will start by hypothesizing that science and metaphysics are not in conflict but that they should join forces in exploring both the visible and the invisible aspects of reality. This epistemic stance had been supported by Karl Popper since 1959 when he stated that "purely metaphysical ideas – and therefore philosophical ideas – have been of the greatest importance for cosmology" (Popper, 1968). Popper considered that "all science is cosmology", as science is expected to solve the problem of cosmology defined as *"the problem of understanding the world – including ourselves, and our knowledge, as part of the world"* (Popper, 1968; emphasis in original). The same understanding of the epistemic role of metaphysics has started to gain acceptance among philosophers of science. For instance, in 1998, Jonathan Lowe defined metaphysics "as the most fundamental form of rational inquiry", one "charting the possibilities of existence" (Lowe, 1998) by going "deeper than any merely empirical science, even physics, because it provides the very framework within which such sciences are conceived and related to one another" (Lowe, 2006). Likewise, Matteo Morganti considers that metaphysics is "an intelligible and autonomous enterprise worth pursuing" in the study of "unobservable entities", as it is "an enterprise dealing with possibilities that – so to put it – are not in competition with those identified with the sciences but, rather, are more general than these and, therefore, play an essential preliminary role with respect to them" (Morganti, 2015).

If we define reality as "the stratified reality", as described by critical realism (Bhaskar, 1975), we can dare to venture into the field of metaphysics. According to critical realism, the world is stratified ontologically in three distinctive realms, the "real", the "actual", and the "empirical", with the "real" domain encompassing the "actual" and the "empirical" domains, but also including various powers (natural mechanisms) and their potentialities (Bhaskar, 1975), such as the folding ability of proteins, the miraculous properties of entanglement in quantum fields, and the ability of the human mind to think original thoughts. The unseen powers, the tendencies and generative mechanisms of the "real" reality which Bhaskar (1975) called "the structured and intransitive objects of scientific inquiry", should be taken as prime realities. In contrast with the intransitive objects of inquiry, the "transitive" objects of knowledge consist of prior existing knowledge which is historically, socially, and culturally informed, and is used by researchers to produce new knowledge. One of the "transitive" objects of knowledge is human faith, or the true belief of "a devout person" which "is not only a doctrine, believed content, but a source of energy that the person who has a faith taps performatively and thus nurtures his or her entire life" (Habermas, 2006). Take, in contrast, the opposing mindset, such as the one of evolutionary biologist Richard Dawkins who believed that "Faith is the great cop-out, the great excuse to evade the need to think and evaluate evidence. Faith is belief in spite of, even perhaps because of, the lack of evidence" (McGrath, 2006). This viewpoint demonstrates a limited understanding of how the human mind functions and how complex thinking processes inform human behaviour. A definition of Christian faith, which can be extended to other theistic religions, explains that faith is not based on irrational lack of evidence, but on "adequate evidence" that the faithful person gathers in order to understand the world. According to this definition, faith

> commences with the conviction of the mind based on adequate evidence; it continues in the confidence of the heart or emotions based on conviction, and it is crowned in the consent of the will, by means of which the conviction and confidence are expressed in conduct.
>
> (Griffith Thomas, 1930)

This epistemic mechanism can be applied not only to religious faith but also to hard-to-explain aspects of reality, which require "interpretation" of visible and invisible facts. This view is shared by mathematical physicist Roger Penrose who believes that the current "faith" in the quantum-mechanics explanation of how matter and energy behave, according to the EPR effect discovered by Einstein, Podolsky, and Rosen in 1935, is unquestionable, as it "has now been observed in the entanglement of particles separated by up to 143 km" (Ball, 2017). However, Penrose questions the "Copenhagen interpretation" of quantum mechanics, named to honour the Danish physicist Niels Bohr (1885–1962). Bohr and his team believed that in quantum-mechanics experiments, such as the two-slit experiment, we cannot know whether light or the electrons behave like a wave or as a particle until we decide to measure them, implying that "the collapse of the wave function" is caused by the observer making the experiment. Penrose rightly considers that the Copenhagen interpretation of quantum mechanics "does not assign any kind of ontological reality to the wave function of a particle, treating it only as a calculational tool for giving us probabilities, which are, in fact, in spectacular agreement with "'real' measurements" (Ball, 2017). This troubles Penrose, who argues that the agreement between theoretical probability and real measurements is enough evidence for conferring "real" ontological status to the quantum state itself (Penrose, 2016). Thus, Penrose recommends a review of our quantum "faith" in light of new evidence, proving that science is always evolving in light of new discoveries of our minds. Penrose is a realist thinker who conceives reality as being made up of three independent and unconnected fields, very similar to Bhaskar's three ontological realms: the field of abstract objects studied by mathematicians; the field of physical reality studied by natural sciences, such as astrophysics (the Universe of space and time, and of matter and energy); and the field of the mind, or of human thought, which, based on Gödel's incompleteness theorem, is "incomputable", as it cannot be captured by a "knowably sound" formal system (Penrose, 1994). I believe that there are obvious relationships between the three fields identified by Penrose, as the mathematical field minutiously describes the physical field, and the mental field can produce physical consequences in the physical field, while the mental field is able to grasp the abstract entities of the mathematical field. If one could conceive that an overarching, unlimited, and purposive Mind runs and unites the three fields of reality, as the author of the language of mathematics, the creator of the ordered, diverse and beautiful physical Universe, and the designer of our creative and morally imbued minds and hearts, we could logically solve the mystery of the Sustainer. Turning to the inference method called abductive reasoning, described by philosopher Charles Sanders Peirce as "reasoning *a posteriori* to a physical hypothesis, or inference of a cause from its effect" (Peirce, 1931–1935), we can. Peirce also called this reasoning method "retroduction" – that is, reasoning "backwards" from consequent to antecedent.

We can use abductive reasoning for trying to decipher the mystery of the life-sustaining system we observe in nature, as abductive reasoning starts from observable facts (there is an artfully crafted and functioning life-sustaining system on Earth) and consists in developing plausible hypotheses or a theory that can explain the observed facts, by making inferences about plausible causes (natural processes, or an intelligent mind), and finally choosing the best explanation. An abductive reasoning syllogism can take the following form:

The surprising fact C is observed;
But if A were true, C would be a matter of course,
Hence, there is reason to suspect that A is true.

(Niiniluoto, 1999)

Even if abductive inferences may not lead to conclusive explanations, as there can be more causes that explain the same effect, the value of this type of reasoning consists in its potential to discover causes of phenomena and to formulate new theories, which cannot happen with a deductive reasoning only process. Peirce's abductive reasoning method has been improved by philosophers of science, such as Gilbert Harman (1965), or Peter Lipton (1991), who developed a way of reasoning called "inference to the best explanation" aiming to solve the problem of causal adequacy of multiple hypotheses in explaining the past. With "inference to the best explanation", scientists compare competing hypotheses with different explanatory powers, and eliminate hypotheses until they can reasonably choose the one that provides the best explanation, which can then be tested to determine its truth validity. One way to assess the truth validity of "the best explanation" is based in probability reasoning. The use of probability calculus to test hypotheses is based on the Bayesian theorem named after English mathematician Thomas Bayes (1701–1761). The theorem is a mathematical formula allowing scientists or philosophers of science to calculate conditional (posterior or prior) probabilities of a hypothesis to be true, given the presence of some observed evidence. In the Bayesian analysis, the conditional probability of some evidence E, given that a specific hypothesis H is true, is calculated in the form of this syllogism: "If H were true, then the surprising fact E would be a matter of course" (or "If H were true, then E *would be expected*") (Meyer, 2021). The inference to the best explanation method has been successfully used to assess metaphysical hypotheses of past causes of events. One such example is the choice of the "big bang" theory as the best cosmological explanation for the origin of the Universe. During the 1930–1960 period, physicists and cosmologists developed two competitive cosmological models of the Universe, namely the "big bang" theory, stating that the Universe is dynamic and is expanding, meaning that it had a beginning in the form of a "singularity" or a "primeval atom" (Farrell, 2006) and consequently will have an end, and the "steady-state" theory, maintaining that the Universe is dynamic and it is expanding by addition of new matter. This latter model did not explain where the new matter comes from, but stated that the Universe is infinite in time and space, so it had no beginning. In 1965, the discovery by two physicists, Arno Penzias and Robert Wilson, of the cosmic background radiation resulting from the Universe's explosive beginning approximately 13.8 billion years ago, discredited the "steady-state" theory and confirmed the stronger explanatory power of the "big bang" theory which is now accepted as the primary valid explanation for the origin of the Universe. Inferring past causes from present evidence is difficult when there are numerous possible causes that can produce a past event. It is easier when scientists can discover an effect for which they can confidently say that there is just one plausible necessary cause (Sober, 1988). The inference that the unique cause C is acting to produce the observed effect E can be expressed in the logical syllogism:

> Cause C is necessary to the occurrence of effect E.
> Effect E exists.
> Therefore, cause C exists.

We can restate this syllogism in the form:

A Sustainer is a necessary cause to the occurrence and functioning of a life-sustaining system.
A life-sustaining system exists.
Therefore, a Sustainer exists.

This logical exercise does not tell us much about the plausible Sustainer, the nature of the life-sustaining system, or the origin of life as ultimate reality. We need to formulate competitive hypotheses about the Sustainer, which can either reject or confirm the universal design intuition. For this I will use the framework of religious faiths as identified by Johann Galtung (1997–1998). Galtung refers to three major religions, namely: "Occidental religions, inspired by the Book, the Old Testament; Hindu religions, called Hinduism, following the tradition of lumping many religious approaches together under one heading; and Oriental religions, inspired by the teachings of the Lord Buddha." Among these diverse religions, Galtung notes how "Life goals change dramatically: from an eternal continuation of individual existence, next to God, to transcendence to a higher existence devoid of individual and permanent identity, *nibban*" (Galtung, 1997–1998), or *nirvana* in Sanskrit, meaning "becoming extinguished" or "blowing out" (Britannica, n.d.).

Distinguishing between a religion with a God who "chooses people", religions with many gods "who are chosen by people", and religions with no god at all, we can identify four worldviews giving different answers to the "cosmic authority problem" (Nagel, 1997): naturalism (no god at all), pantheism (an impersonal deity present throughout the world), deism (a personal, intelligent, transcendental God who does not act within creation), and theism (a personal, intelligent, transcendental, and immanent God permanently acting within creation). For our analysis, I will compare only the two clear-cut hypotheses, the "naturalism" hypothesis and the "theism" hypothesis, as the worldviews which can help answer our research question.

Naturalism is a materialistic worldview, which assumes that matter and energy are the ultimate realities in the Universe, itself a closed system of causes and effects not impacted by external forces. Those who hold this worldview believe "God to be a product of human imagination, which they believe to be a product of material evolution" (Axe, 2016). They do not see intentional design in the Universe, but only "blind physical forces and genetic replication", including in the evolution of life. In the words of biologist Richard Dawkins:

In a universe of electrons and selfish genes, blind physical forces and genetic replication, some people are going to get hurt, other people are going to get lucky, and you won't find any rhyme or reason in it, nor any justice. The universe that we observe has precisely the properties we should expect if there is, at bottom, no design, no purpose, no evil, no good, nothing but pitiless indifference.

Dawkins (1995)

We can express Dawkins' naturalism in the following syllogism, as an abductive inference:

Logic: If "blind, pitiless" matter and energy rather than a Mind is the prime reality from which all else originated, then we would expect no evidence of intelligent design in life and the universe, rather only evidence of apparent design.
Data: Life and the universe do not exhibit evidence of actual design, only apparent design.
Conclusion: Therefore, we have reason to believe that life and the universe are the product of blind materialistic forces rather than a preexisting Mind.

Meyer (2021)

Theism is the worldview of the three major world religions, Jewish, Christian, and Muslim. This worldview asserts the priority of an omniscient Mind over matter. The world and everything in it has been brought into existence by a nonmaterial, intelligent cosmic being, an infinite, transcendental Mind, who not only created it but also acts in the Universe sustaining it through natural laws and other unexplained occurrences. In order to find epistemic support for a Sustainer, we can develop the following syllogism, starting from the "universal design intuition" (Axe, 2016):

Logic: If scientific-metaphysical facts are in harmony with the universal design intuition, we would expect that our Universe would be carefully organized to support life on Earth.
Data: We have surprising evidence that a life-sustaining system exists and functions, supporting life on Earth in many different forms, including conscious life.
Conclusion: Therefore, we have reason to believe that the life-support system is the product of a powerful, infinitely intelligent, out-of-this world Being, who runs and sustains it.

Scientific-metaphysical evidence that the world exists and has an ordered and intelligible causal structure has been provided in previous chapters of this book, discussing the origin of the Universe, a process which presupposes a vastly knowledgeable initiator, and the apparent fine-tunning of the Earth to support life, which indicates design. There is also evidence, provided since 1953, when James Watson and Francis Crick discovered the information bearing-structure of the deoxyribonucleic acid (DNA) molecule and its role not only in transmitting hereditary information but also in providing precise information for building protein molecules, that "the ultimate stuff of life" (Denton, 1986) is not the result of chance but of an intelligent Mind. Naturalistic theories about the origin of life have started to be challenged for failing to provide a credible explanation for the origin of the genetic information necessary to produce the first living cell. The explanations provided by these theories for the origin of life were based on either chance (Monod, 1971), the laws of physics and chemistry (Crick, 1981; Dawkins, 1995), or on a combination of chance and natural laws (de Duve, 1995). In 1984, in the book *The Mystery of Life's Origin* (Thaxton, Bradley and Olsen, 1984), a trio of interdisciplinary scientists with backgrounds in chemistry, materials science, and geochemistry advanced the hypothesis that an intelligent cause should be considered a legitimate scientific explanation for the origin of the code of life in the DNA molecule: "You can't get gold out of copper, apples out of oranges, or information out of negative thermal entropy." The same "seemingly intractable difficulty of explaining how a living system could have gradually arisen as a result of known chemical and physical processes" was also identified in 1985 by the Australian microbiologist Michael Denton in his book *Evolution: A Theory in Crisis*, in which he explained that the difficulty "raises the obvious possibility that factors as yet undefined by science may have played some role" (Denton, 1986). The completion of the Human Genome Project in 2000, dubbed the "Book of Life" (Pennisi, 2000), resuscitated interest in the hypothesis of design of life by an intelligent agent. Books by Stephen Meyer (2009, 2013, 2021) and Douglas Axe (2016) provided evidence from the fossil record and from microbiology that building new life forms requires new sources of specific genetic or biological information which cannot be the result of unguided evolution, as only "intelligent agents routinely produce vast amounts of specified information in order to communicate and to build a variety of new structures" (Meyer, 2021). The explanatory deficits of Neo-Darwinism in origin-of-life issues have also been acknowledged in the last three decades by evolutionary biologists, such as those in the Altenberg 16 group who proposed an extended evolutionary synthesis (Muller, 2017).

As for the epistemic acceptance of the scientific-metaphysical explanations concerning ultimate and origin-of-life issues, it depends on anyone's stance, provided that it is "internally consistent and coherent": "whether or not one takes metaphysics as an enterprise worth pursuing is just a matter of choosing one's 'stance'" (Chakravartty, 2007). This choice has immense significance, since it can lead to two understandings of life which cannot be both true. One understanding can see life as "mystery and masterpiece – an overflowing abundance of perfect compositions" (Axe, 2016) with numerous living creatures being together, among whom only humans have been given the privilege to discern, to understand, and to wonder at the immensity and creativity of the life project to which they are invited to contribute. Even if humans do not understand every aspect of the life project, by participating in it they are part of the wholeness, and can be hopeful, as they have "eternity imprinted in their hearts" (Ecclesiastes 3:11). An inference that only an intentional, divine Intelligence, higher than created humans and able to confer meaning to the creation, can be the Author and Sustainer of it all, is a matter of course, for humans with "mental incandescence" (Rolston, 1989) who are "conscious thinkers with a moral sense" (Axe, 2016). I have found that by choosing this stance, life becomes both more bearable, as it takes place within physical and moral limits I know and accept, and more fulfilling, as this partnership for life experience infuses me with a sense of self-worth and of what is really true. The other stance sees life as a cosmic accident, "the outcome of accidental colocation of atoms" (Russell, 1957), or as a process run by a personified mechanism of natural selection whose only goal is genetic replication but which can never reach perfection as, according to Darwin, it is

> daily and hourly scrutinizing, throughout the world, the slightest varia-
> tions; rejecting those that are bad, preserving and adding up all that are
> good; silently and insensibly working, whenever and wherever opportu-
> nity offers, at the improvement of each organic being in relation to its
> organic and inorganic conditions of life.
>
> (Darwin, 1859)

The problem with the theory of unguided evolution is that it assigns creative powers to a series of unthinking processes involving physical things, and excludes itself from the vast realm of the mental where new knowledge about new life is purposefully created by non-physical thoughts, consciousness, reason, and values, all attributes of creative minds belonging to real thinking people. This theory supresses the design intuition, and makes the search for the truth about the most significant aspects of existence, including life itself, futile. Those people who accept this stance are left alone in a strange Universe where whatever meaning, purpose, or joy there is must be produced by themselves (Krauss, 2017). Being more "internally consistent and coherent" (Chakravartty, 2007), the theistic stance gives more reasons to accept the universal design intuition hypothesis which, in my mind, has stronger explanatory power than the evolutionary naturalistic hypothesis to answer this chapter's research question: To whom do we owe the existence of the life-sustaining web we see freely available to all living earthly creatures? With a relatively limited mind, one cannot aim to describe the indescribable attributes of the all-powerful, omniscient, and generous Sustainer, whose thoughts and ways are higher than the thoughts and ways of humans. However, we can perceive these attributes from the facts obvious in nature which give meaning to human existence, and help shape humans' character and deeds, preparing them for the adventure of sustainability, whenever humans choose to join in the work of the Sustainer.

This chapter has argued for the need to reconsider metaphysics as an important source of knowledge about reality. It has used the stratified reality concept of critical realism and the method of inductive reasoning called "inference to the best explanation" to search for the most plausible cause of the life-sustaining web of relationships in which humans live. Comparing the worldview of naturalism or scientific materialism with the worldview of theism and the existing evidence about the origin of the Universe, the fine-tuning of the Universe to support life on Earth, and the facts provided by genome analysis in biology proving the complex design in life-supporting protein molecules, it has concluded that theism offers a better explanation about the cause behind the complex, dynamic, and diverse life-sustaining network than scientific materialism.

References

Axe, D. (2016) *Undeniable How Biology Confirms Our Intuition That Life Is Designed.* New York: HarperOne.

Ball, S. (2017) "Review of book "Fashion, Faith, and Fantasy in the New Physics of the Universe" by Roger Penrose." *Perspectives on Science and Christian Faith*, 69(3). www.asa3.org/ASA/PSCF/2017/PSCF9-17Complete.pdf.

Barbour, I.G. (1997) *Religion and Science: Historical and Contemporary Issues.* San Francisco: HarperSanFrancisco.

Berger, P.L. (ed.) (1999) *The Desecularization of the World.* Washington, DC: Ethics and Public Policy Center.

Bhaskar, R. (1975) *A Realist Theory of Science.* Leeds: Leeds Books.

Boyle, R. (1686) *A free enquiry into the vulgarly received notion of nature*, edited by E. B. Davis and M. Hunter. Cambridge University Press.

Britannica, The Editors of Encyclopaedia (n.d.) "Nirvana." Encyclopedia Britannica. www.britannica.com/topic/nirvana-religion.

Brubaker, R. (2017) "Between nationalism and civilizationism: The European populist moment in comparative perspective." *Ethnic and Racial Studies*, 40: 1191–1226.

Butterfield, H. (1957) *The Origins of Modern Science.* New York: Free Press.

Chakravartty, A. (2007) *A Metaphysics for Scientific Realism: Knowing the Unobservable.* Cambridge: Cambridge University Press.

Copleston, F. (1993) *A History of Philosophy*, Vol. 3. New York: Doubleday.

Crick, F. (1981) *Life Itself: Its Origins and Nature.* New York: Simon & Schuster.

de Duve, C. (1995) *Vital Dust: Life as a Cosmic Imperative.* New York: Basic Books.

Daly, H.E. and Farley, J. (2011) *Ecological Economics Principles and Applications.* Washington, DC: Island Press.

Davis, T. (2013) "The Faith of a Great Scientist: Robert Boyle's Religious Life, Attitudes, and Vocation" *BioLogos*, August 8, 2013. https://biologos.org/articles/the-faith-of-a-great-scientist-robert-boyles-religious-life-attitudes-and-vocation.

Dawkins, R. (1986) *The Blind Watchmaker: Why the Evidence of Evolution Reveals a Universe Without Design.* New York: Norton.

Dawkins, R. (1995) *River Out of Eden: A Darwinian View of Life.* New York: Basic Books.

Darwin, C. (1859) *On the Origin of Species by Means of Natural Selection*, 1st edition. London: John Murray.

Denton, M. (1986) *Evolution: A Theory in Crisis.* Bethesda, MD: Adler and Adler.

Denton, M. (2022) *The Miracle of Man: The Fine Tuning of Nature for Human Existence.* Seattle, WA: Discovery Institute Press.

Draper, J.W. (1874) *History of the Conflict Between Religion and Science.* New York: Appleton.

Durán, A.J. (2019) "Laplace, Napoleon, and God." Blog del Instituto de Matematicas de la Universidad de Sevilla, 18 February 2019. https://institucional.us.es/blogimus/en/2019/02/laplace-napoleon-and-god.

Dyson, F. (2006) "Religion from the Outside." *The New York Times Review of Books*, 53(11). https://image.sciencenet.cn/olddata/kexue.com.cn/blog/admin/images/upfiles/200791262443795967.pdf.

Eaton, W. (n.d.) "Robert Boyle (1627–1691)." Internet Encyclopedia of Philosophy. https://iep.utm.edu/robert-boyle.

Farrell, J. (2006) *The Day Without Yesterday: Lemaitre, Einstein, and the Birth of Modern Cosmology.* New York: Basic Books.

Fuller, S. (2007) *Science vs. Religion? Intelligent Design and the Problem of Evolution.* Oxford: Polity.

Galtung, J. (1997–1998) "Religions: Hard and Soft." *CrossCurrents*, 47(4): 437–450.

Gopnik, A. (2014) "See Jane Evolve: Picture Books Explain Darwin." Mind and Matter, *Wall Street Journal.* http://alisongopnik.com/Alison_Gopnik_WSJcolumns.htm#18Apr14.

Grant, C.D. (1984) *God the Center of Value: Value Theory in the Theology of H. Richard Niebuhr.* Fort Worth: Texas Christian University Press.

Griffith Thomas, W.H. (1930) *The Principles of Theology: An Introduction to the Thirty-Nine Articles.* Harlow, UK: Longmans, Green and Co.

Habermas, J. (2006) "Religion in the Public Sphere." *European Journal of Philosophy*, 14(1): 1–25.

Harman, G. (1965) "Inference to the Best Explanation." *The Philosophical Review*, 74: 88–95.

Hawking, S. (1998) *A Brief History of Time: From the Big Bang to the Black Holes.* New York: Bantam Books.

Hawking, S. and Mlodinow, L. (2010) *The Grand Design.* London: Bantam Books.

Haynes, J. (2005) "Religion and international relations after '9/11'." *Democratization*, 12: 398–413.

Hodgson, P.E. (2001) "The Christian Origin of Science." *Logos: A Journal of Catholic Thought and Culture*, 4(2): 138–159.

Hooykaas, R. (1972) *Religion and the Rise of Modern Science.* Grand Rapids, MI: Eerdmans.

Hume, D. (1989) *Dialogues Concerning Natural Religion.* Buffalo, NY: Prometheus.

Huntington, S.P. (1993) "The Clash of Civilizations?" *Foreign Affairs*, 72(3): 22–49.

Inglehart, R.F. (2018) *Cultural Evolution People's Motivations are Changing, and Reshaping the World.* Cambridge: Cambridge University Press.

Jerath, R. and Beveridge, C. (2018) "Top Mysteries of the Mind: Insights from the Default Space Model of Consciousness." *Frontiers in Human Neuroscience*, 12 (2018). https://doi.org/10.3389/fnhum.2018.00162.

Kahneman, D. (2011) *Thinking, Fast and Slow.* New York: Farrar, Straus and Giroux.

Kant, I. (1781) *The Critique of Pure Reason*, translated by J.M.D. Meiklejohn. London: Henry G. Bohn.

Kepel, G. (1994) *The Revenge of God: The Resurgence of Islam, Christianity and Judaism in the Modern World.* University Park, PA: Pennsylvania State University Press.

Koyre, A. (1957) *From the Closed World to the Infinite Universe.* New York: Harper & Brothers.

Krauss, L. (2017) "The Universe Doesn't Give a Damn About Us." *Real Clear Science*, April 21, 2017. www.realclearscience.com/video/2017/04/21/the_universe_doesnt_give_a_damn_about_us.html?mobile_redirect=true.

Lipton, P. (1991) *Inference to the Best Explanation*. London: Routledge.

Lowe, E.J. (1998) *Possibility of Metaphysics: Substance, Identity, and Time*. Oxford: Oxford University Press.

Lowe, E.J. (2006) *The Four-Category Ontology*. Oxford: Oxford University Press.

MacIntyre, A.C. (1981) *After Virtue: A Study in Moral Theory*. Notre Dame, IN: University of Notre Dame Press.

Marenbon, J. (2023) "Medieval Philosophy." *The Stanford Encyclopedia of Philosophy* (Spring 2023 Edition), edited by E.N. Zalta and U. Nodelman. https://plato.stanford.edu/archives/spr2023/entries/medieval-philosophy.

Marx, K. (1970) "Introduction." In *A Contribution to the Critique of Hegel's Philosophy of Right*, translated by A. Jolin and J. O'Malley, edited by J. O'Malley. Cambridge: Cambridge University Press. (Original work published 1843)

McGilchrist, I. (2021) *The Matter with Things: Our Brains, Our Delusions and the Unmaking of the World*. Perspectiva Press.

McGrath, A. (2006) *Christianity, An Introduction*. Hoboken, NJ: John Wiley & Sons.

Medawar, P.B. (1984) *The Limits of Science*. New York: Harper & Row.

Meyer, S. (2009) *Signature in the Cell*. New York: HarperOne.

Meyer, S. (2013) *Darwin's Doubt: The Explosive Origin of Animal Life and the Case for Intelligent Design*. San Francisco: HarperOne.

Meyer, S. (2021) *Return of the God Hypothesis. Three Scientific Discoveries that Reveal the Mind Behind the Universe*. New York: HarperOne.

Monod, J. (1971) *Chance and Necessity: An Essay on the Natural Philosophy of Modern Biology*. New York: Vintage.

Morganti, M. (2015) "Science-based Metaphysics: On Some Recent Anti-Metaphysical Claims." *Philosophia Scientiæ*, 19(1): 57–70. http://philosophiascientiae.revues.org/1038.

Muller, G.B. (2017) "Why an extended evolutionary synthesis is necessary." *Interface Focus*, 7: 20170015. http://dx.doi.org/10.1098/rsfs.2017.0015.

Nagel, T. (1997) *The Last Word*. Oxford: Oxford University Press.

Niiniluoto, I. (1999) "Defending Abduction." *Philosophy of Science*, 66, Supplement. Proceedings of the 1998 Biennial Meetings of the Philosophy of Science Association. Part I: Contributed Papers: S436–S451.

Nye, J.S. (1990) "Soft power." *Foreign Policy*, 80: 153–171.

Oakley, F. (1961) "Christian Theology and the Newtonian Science: The Rise of the Concept of the Laws of Nature." *Church History*, 30(4): 433–457.

Ozturk, A.E. (2023) "Religious Soft Power: Definition(s), Limits and Usage." *Religions*, 14(2): 135. https://doi.org/10.3390/rel14020135.

Peirce, C.S. (1931–1935, 1958) *Collected Papers*, Volumes 1–6, edited by C. Hartshorne and P. Weiss; Volumes 7–8, edited by A. Burks. Cambridge, MA: Harvard University Press.

Pennisi, E. (2000) "Finally, the Book of Life and Instructions for Navigating It." *Science*, 288(5475): 2304–2307.

Penrose, R. (1994) *Shadows of the Mind*. New York: Oxford University Press.

Penrose, R. (2016) *Fashion, Faith and Fantasy in the New Physics of the Universe*. Princeton, NJ: Princeton University Press.

Popper, K. R. (1968) *The Logic of Scientific Discovery*. London: Hutchinson and Co. First published in 1959.

Popper, K. (1978) "Natural Selection and the Emergence of Mind." *Dialectica*, 32(3–4): 339–355.

Rolston, H. III (1989) *Philosophy Gone Wild: Essays in Environmental Ethics.* Amherst, NY: Prometheus.

Russell, B. (1957) *Why I Am Not a Christian*, edited by P. Edwards. New York: Simon and Schuster.

Sacks, J. (2011) *The Great Partnership: Science, Religion, and the Search for Meaning.* NewYork: Schocken Books.

Sober, E. (1988) *Reconstructing the Past*. Cambridge, MA: MIT Press.

Thaxton, C., Bradley, W.L., and Olsen, R.L. (1984) *The Mystery of Life's Origin: Reassessing Current Theories.* New York: Philosophical Library.

Whitehead, A.N. (1925) *Science and the Modern World*. New York: Free Press.

Wigner, E. (1960) "The Unreasonable Effectiveness of Mathematics." *Communications on Pure and Applied Mathematics*, 13(1): 1–14. https://doi.org/10.1002/cpa.3160130102.

Conclusion

This book was written from a desire to discuss sustainability not only as a concept but as an objectively existing reality functioning as a miraculous life-sustaining system which has secured continuation of life on planet Earth for more than three billion years (Marshall, 2023). This approach to sustainability implies that life in all forms is precious, as it is worth sustaining, and that humans, as the only living beings on Earth endowed with reason and "the moral law within" (Kant, 1788), are expected to participate responsibly in the life-web at the core of sustainability, obeying the pre-set rules of functioning of the physical Universe. In other words, more is expected from humans than from a frog, a spider, an elephant, or any other non-rational creature.

I first understood the wisdom of such an approach to sustainability while doing research in Costa Rica in 2016. There was a note on the hotel room door which stated: "Dear Guests, in this room you may find insects. Please do not kill them, call someone from the reception who will deal gently with them. They belong to our rich biodiversity, we love them and want to protect them." What a reality check! My cultural upbringing to that point had told me that insects daring to bother or inconvenience humans in any way are to be eliminated, full stop. The hotel door note made me think more deeply about how we humans do not own the biosphere, which belongs as much to insects, to microorganisms, to mammals, and to all other forms of life. Insects do not encroach on our territory; perhaps we humans crowd them out by continuously expanding our material culture through buildings, highways, and other built infrastructures, without fully understanding how we harm them and prevent them from fulfilling the roles they play in our broader ecosystems.

Defining sustainability ontologically has allowed me to consider some of the big questions about the meaning of life on Earth for humans, such as how the life-sustaining aggregate came to be on our planet; to what or to whom do we owe our continuous existence as living, thinking, loving, and hating beings; and, ultimately, what the future of life on Earth may be. I have also tried to assess answers humans have already given to these questions along humanity's history, mainly in light of the consequences we can now see all around us and which are hard to deny: nature despoiled, and global

DOI: 10.4324/9781003307587-10

biodiversity loss "potentially ushering in the sixth mass extinction" (Harfoot et al., 2021), due to threats posed by human activities such as industrial farming, hunting and trapping, logging, over-polluting, and climate change; a food crisis building up; an uncontrolled rise in inequality, violence, and mass displacement and migration, and a loss of meaning, desperation, and little appreciation for life, including human life. While not offering a clear-cut solution to humanity's unsustainability problems, the book highlights the need for change, discussing some needed actions.

There are three conclusions I would like to leave readers with. First, sustainability is poorly understood due to the mistaken idea (i.e. the epistemic fallacy) that "the map is able to correctly explain the territory", or the faulty assumption that questions about existence can only be framed in terms of people's knowledge or "concepts" (Bhaskar, 2008). By focusing on the *concept* of sustainability, which has up to now been defined in strong interconnection with "sustainable development" as a concept, we have failed to thoroughly explore sustainability as it occurs in ecosystems – where life happens both in nature's biotic communities and in human societies, which are strongly interconnected. As a result, we have lost the idea of biotic unity or the value of life in all its forms, and have a poor understanding of the structures, functions, processes, and relationships in nature which seem specifically designed to protect and sustain life. Some are visible, such as birds migrating to warmer places before winter begins, or bioluminescent adult fireflies glowing in order to attract mates, while others are invisible, such as photosynthesis, metabolism, or the immune system. There is no agreement among scientists about how many species currently exist in the Earth's biosphere. A 2011 study reported about 8.7 million species, but it is believed that more than 80 per cent of terrestrial species and more than 90 per cent of marine species are still undiscovered (Sweetlove, 2011). More studies are needed to better understand the various terrestrial and aquatic ecosystems' structure, functions, and processes, as well as their operation and interrelationships, and the boundary conditions for securing reproduction of their living species, *in their own right*, not only in light of their usefulness to us as human beings. Such studies should clearly offer clues of the true objective limits to human economic activities and help humans reignite a lost sense of wonder for the gift of life on our miraculously life-friendly planet.

This leads me to the second conclusion. The modernist faith in science as the only tool for exploring and finding causes or mechanisms acting to sustain life – both in nature and in society – has led to a simplified image of both nature and human nature. Science has no theory of how the first living cell came to be. Biochemists know the basic classes of chemical compounds of any living cell (nucleotides, containing genetic information, carbohydrates, amino acids or proteins, and lipids), but they do not know how and why their integration produces the spark of life. It is an important task for biologists to go beyond producing synthetic cells to producing credible explanations about the origin of the information stored in the DNA molecule of the living cell.

Likewise, behavioural sciences currently paint a deficient picture of humans acting as if there is no difference between the material substance of their brains and the immaterial content of the mind – the conceptual realm where thinking occurs, and where imagination and intuitions, leading to creative thinking, are born. Finding answers to the deep questions concerning life in nature and in human societies requires both rigorous scientific investigation and an open philosophical mind in interpreting the research findings, by acknowledging "the importance of the more fundamental realities that lend meaning to science" (Axe, 2016). This book uses critical realism (Bhaskar, 1975, 1979, 2008, 2020) as its theoretical grounding. Critical realism is a realist philosophy best able to explain sustainability by connecting the ontological realism of natural sciences empirically observing nature in order to discover its law-like regularities (Oakley, 1961) with the epistemological relativism and judgmental rationality of social sciences exploring how humans relate to sustainability. According to critical realism, (1) a world outside and independent of our conscious perception exists, with only some aspects of this world being objectively knowable (empirically) via our senses; (2) reality is stratified in three domains: the empirical (what we experience), the actual (what science reveals), and the real (what science can explain). *Empirically*, we know that we live in a life-friendly Universe where we have air to breathe, potable water, waterways for travel, starry skies at night, rainforests teeming with biodiversity, and soils to grow crops. Scientific inquiry can reveal *actual reality*. According to biometric dating of terrestrial rock formations, the Earth has existed for approximately 4.55 billion years (Patterson, 1956) and has provided life-sustenance in the earliest habitable environments, the submarine-hydrothermal vents, for more than 3,770 million years (Dodd et al., 2017). Science also tells us that *Homo sapiens* has been around for about 286,000 years (Richter et al., 2017). Understanding the *real domain* is the proper role of science, namely to develop theories able to explain why certain events occur, or what the generative mechanisms that lie beyond the empirical domain might be. For instance, why do standing trees sequester carbon? Or why do Monarch butterflies feed on milkweed? Why do people fall in love? To explain the nature and workings of a generative mechanism, deductive reasoning, as noted by Karl Popper's theory of falsifiability (Popper, 1992), is insufficient. Critical realist researchers need to use also inductive reasoning methods, such as retroduction or abduction, to find the best explanation of an observed phenomenon, by reasoning about alternative possible hypotheses. Contrary to deductive reasoning, where an existing theory is used to explain a particular event, abductive reasoning has the potential to create new theoretical knowledge starting from observable facts. Using abductive reasoning, the discovery of "quantum resources", honoured by the 2022 Nobel prize in physics, was possible. The method also allowed me to hypothesize about who an "out-of-this-world" Originator and Sustainer of the life-sustaining mechanism on our planet might be.

The third conclusion is that sustainability efforts should not be focused at the level of a whole society, which is an aggregate "unthinking" entity, but at the level of alive individual human beings who, when "standing on the shoulders of the giants" in science, philosophy, religion, art, sports, and music, and using their own rational and moral abilities, are able to discover the truth about life on Earth and the real meaning of sustainability. It is my belief that any human being who understands that sustainability is not a human concept but a real gift of life can learn how to live sustainably, by making not only "useful" choices for themselves but also "right" choices for other living beings, including their neighbours. Once freed from the shackles of their "egos", any human being can become a sustainable person and a meaningful participant in the cosmic dance of the living, whose purpose is not only survival but living life to the fullest, for all involved, thus fulfilling the potential of their real selves for the benefit of this and future generations. Then the meaning found in wisdom books may start to make sense, making humans glad to have their intuitions confirmed that we are here not by chance, we are valuable and loved, and we have a future: "He has made everything beautiful in its time. He has also set eternity in the human heart; yet no one can fathom what God has done from beginning to end" (Ecclesiastes 3:11).

References

Axe, D. (2016) *Undeniable How Biology Confirms Our Intuition That Life Is Designed.* New York: HarperOne.

Bhaskar, R. (1975) *A Realist Theory of Science.* London: Routledge.

Bhaskar, R. (1979) *The Possibility of Naturalism: A Philosophical Critique of the Contemporary Human Sciences.* Atlantic Highlands, NJ: Humanities Press.

Bhaskar, R. (2008) *Dialectic: The Pulse of Freedom.* London: Routledge.

Bhaskar, R. (2020) "Critical realism and the ontology of persons." *Journal of Critical Realism* 19(2): 113–120. https://doi.org/10.1080/14767430.2020.1734736.

Dodd, M.S., Papineau, D., Grenne, T.*et al.* (2017) "Evidence for early life in Earth's oldest hydrothermal vent precipitates." *Nature,* 543(7643): 60–64. https://doi.org/10.1038/nature21377.

Harfoot, M.B.J., Johnston, A., Balmford, A., Burgess, N.D., Butchart, S.H.M., Dias, M.P., Hazin, C., Hilton-Taylor, C., Hoffmann M., Isaac, N.J.B., Iversen, L.L., Outhwaite, C.L., Visconti, P. and Geldmann, J. (2021) "Using the IUCN Red List to map threats to terrestrial vertebrates at global scale." *Nature, Ecology & Evolution,* 5: 1510–1519www.nature.com/natecolevol.

Kant, I. (2007) "Critique of Practical Reason." in *The Cambridge Companion to Kant and Modern Philosophy,* edited by P. Guyer. (Original work published 1788). https://doi.org/10.1017/CCOL052182303X.

Marshall, M. (2023) "Timeline: The Evolution of Life." *New Scientist,* 27 April 2023. www.newscientist.com/article/dn17453-timeline-the-evolution-of-life.

Oakley, F. (1961) "Christian Theology and the Newtonian Science: The Rise of the Concept of the Laws of Nature." *Church History,* 30(4): 433–457.

Patterson, C.C. (1956) "Age of meteorites and the earth." *Geochimica and Cosmochimica Acta,* 10(4): 230–237.

Popper, K. (1992) *The Logic of Scientific Discovery.* London and New York: Routledge.

Richter, D., Grün, R., Joannes-Boyau, R., Steele, T.E., Amani, F., Rué, M., Fernandes, P., Raynal, J.-P., Geraads, D., Ben-Ncer, A., Hublin, J.-J., and McPherro, S.P. (2017) "The age of the hominin fossils from Jebel Irhoud, Morocco, and the origins of the Middle Stone Age." *Nature*, 546: 293–296.

Sweetlove, L. (2011) "Number of species on Earth tagged at 8.7 million." *Nature*. https://doi.org/10.1038/news.2011.498.

Index

Note: Illustrations are indicated by page numbers in *italics*.

For Product Safety Concerns and Information please contact our EU
representative GPSR@taylorandfrancis.com
Taylor & Francis Verlag GmbH, Kaufingerstraße 24, 80331 München, Germany